Fields and/or Particles

Fields and/or Particles

D. K. SEN
Department of Mathematics,
University of Toronto,
Toronto, Canada

1968

THE RYERSON PRESS
TORONTO

ACADEMIC PRESS
LONDON AND NEW YORK

ACADEMIC PRESS INC. (LONDON) LTD
BERKELEY SQUARE HOUSE
BERKELEY SQUARE
LONDON, W.1

U.S. Edition published by
ACADEMIC PRESS INC.
111 FIFTH AVENUE
NEW YORK, NEW YORK 10003

Library of Congress Catalog Card Number: 67-30767

MADE AND PRINTED IN GREAT BRITAIN BY OFFSET BY
WILLIAM CLOWES AND SONS, LIMITED
LONDON AND BECCLES

Preface

The object of this monograph is to provide a brief *survey* of some of the fundamental theories of physics from an overall viewpoint. The underlying theme here is the problem of field-particle duality. From this viewpoint one can classify the major physical theories into three distinct categories: dualistic, non-dualistic and unified non-dualistic. I have tried to describe the historical attempts to understand and overcome the problem of duality by taking illustrative examples from the well established as well as the *not so well established* theories. Only the essentials are presented and no attempt is made to go into details or give any applications.

In keeping with the underlying theme I have included neither modern axiomatic quantum field theory nor S-matrix and dispersion theory, as these later developments do not shed any new light on the basic problem of duality. For the same reason symmetries are mentioned only as far as they are relevant in understanding Heisenberg's unified field theory.

To graduate students of theoretical physics the book should serve as an introductory text on theories of fields and particles. The mathematical steps are kept deliberately concise; it is hoped that no serious student will have any difficulty in working them out. It was, however, felt necessary to include brief summaries (without proofs) of the fundamental properties of Riemannian geometry and Hilbert spaces. For a detailed understanding of these two disciplines the reader must refer to mathematical texts on the respective topics.

Most of the material presented here has appeared elsewhere in some form or other. The original sources are acknowledged as far as possible; I am also, of course, indebted to existing textbooks, some of which are referred to in the general reference at the end.

I would like to acknowledge here a partial research grant from the National Research Council of Canada which made this project possible. Finally, I am indebted to the editorial staff of the Academic Press for their painstaking cooperation in producing this book.

Toronto, 1968 D. K. SEN

Table of Contents

Introduction

In the history of fundamental physics no other basic concepts have assumed a more important role than that of *fields* and *particles*. In fact it seems to be a strange characteristic of the human mind that it is forced to describe the physical properties of matter either as fields or as particles. The whole history of physics appears as a struggle to either clarify or escape from this *either or* dichotomy.

The concept of particles appeared first and supplemented by the action-at-a-distance principle achieved its zenith in the heyday of Newtonian mechanics. With the advent of electromagnetism and the downfall of the action-at-a-distance principle it was necessary to introduce the field concept as a supplement to the particle concept. But electromagnetism did not replace particles by field; one had to use the two notions side by side. This is clearly demonstrated by the fact that in classical electrodynamics one has two sets of equations: the so-called field equations describing the connection between the field and its source, i.e., charged particles; and the equation of motion which describes the motion of particles under the action of the field.

The connection between the field and its source has always been and still is the most difficult problem in classical and quantum electrodynamics. The concept of point particle sources has led for example to infinite self-energies whereas the notion of extended sources still remains unsatisfactory because of their inevitable arbitrariness.

We shall call a theory *dualistic* if it supposes that the source of the field, i.e., the particles with their characteristic masses and charge, etc., form a separate entity apart from the field which they generate. In this category belongs classical electrodynamics and the general theory of relativity. Although the field-particle duality is inherent in both these theories the latter differs from the former among other things in the following respect: it can dispense with a separate equation of motion; the equation of motion of field-singularities follow from the field equations alone, but the mass or charge of the particles enter into the theory as integration constants.

The advent of quantum theory gave rise to a different type of field-particle duality. Whereas in classical theory the field and particles were separate entities, in quantum theory each particle exhibits its own proper field as a result of indeterminacy of simultaneous measurement of its classical particle characteristics. The quantum field-particle duality should be clearly differentiated from the classical one.

Historically the first attempt to escape from this field-particle dichotomy on the classical level was made by Mie† in 1912. Mie demonstrated that it is

† Mie, G. (1912) *Ann. Physik*. **37**, 511; **39**, 1.

possible to escape from the field-particle duality in electromagnetism and derive the existence of charged particles from a theory of field alone. Mie's ambitious theory, however, was open to among other things the following objection: it depended on the absolute values of potentials in such a way as to prohibit the existence of a material particle in a constant potential field. Born and Infeld† circumvented this difficulty by basing their theory directly on the field variables, the masses and charge of the source particles being completely determined by the characteristics of the field.

We shall call a theory *non-dualistic* if it is based on the concept of a pure field or if it uses only particles as fundamental constituents of matter. On the classical level we have then the non-dualistic pure field theories of Mie and Born-Infeld. But if it is possible to construct a pure field theory it should also be possible to have a pure particle theory. Wheeler and Feynman‡ have revived the action-at-a-distance principle in electrodynamics and have shown that the Maxwell-Lorentz theory of purely retarded interaction between charges plus the self-force for the individual charges describing radiation damping is equivalent to a pure particle theory with half-retarded plus half-advanced interactions between the charges. Immediately after Einstein, Whitehead§ gave an action-at-a-distance theory of gravitation in a flat space in which particles interact through retarded gravitational potentials. Recently Hoyle and Narlikar|| have formulated a pure particle theory of gravitation in which Mach's principle is given a firm theoretical basis, a scheme which eluded even Einstein although he was profoundly influenced by it.

Without any doubt the non-dualistic theories have some highly satisfactory features compared to dualistic ones. First of all there is an economy of concepts, and secondly on the classical level at least they are free from any divergence difficulties.

By far the most satisfactory non-dualistic theory is the quantum theory of interacting fields which follows from a direct extrapolation of quantum theory of one particle to a many particle formalism.

Whereas the classical non-dualistic theories both in particle and field formalisms are free from divergence difficulties quantum field theory is still beset with the same infinity problems which plagued the dualistic Maxwell-Lorentz theory. The renormalization scheme has not succeeded in removing the basic conceptual difficulties of the theory; the masses and coupling constants enter in quantum field theory still as arbitrary parameters.

Physics would thus seem to have finally emerged from the field-particle

† Born, M., and Infeld, L. (1934) *Proc. R. Soc. A.* **144**, 425.
‡ Wheeler, J. A. and Feynman, R. P. (1949) *Rev. mod. Phys.* **21**, 425.
§ Whitehead, A. N. (1922) "The Principle of Relativity." Cambridge.
|| Hoyle, F. and Narlikar, J. V. (1964) *Proc. R. Soc. A.* **282**, 191.

dichotomy albeit in a not too satisfactory manner. It is now time to tackle another problem, that of unification of all fields or particles. This grandiose scheme of constructing a unified non-dualistic theory of matter has been considered on the classical level by Einstein-Schrödinger† and Wheeler-Misner‡ and on the quantum level by Heisenberg.§ Both these theories are field theories, so that the balance would seem to be in favour of fields as a primary concept. But it is quite conceivable that a future satisfactory theory will be based on an entirely different primary concept other than fields or particles.

Summarizing then we have the following table for reference.

Dualistic theories

- Classical
  - Electrodynamics (Maxwell-Lorentz)
  - Gravitation (Newton-Einstein)
- Quantum (theory of a particle)

Non-dualistic theories

- Classical
  - Particle formalism
    - Electrodynamics (Wheeler-Feynman)
    - Gravitation (Whitehead and Hoyle-Narlikar)
  - Field formalism (Electrodynamics of Mie and Born-Infeld)
- Quantum (theory of fields)

Unified non-dualistic theories

- Classical (Einstein-Schrödinger and Wheeler-Misner)
- Quantum (Heisenberg)

† Einstein, A. and Strauss, E. G. (1946) *Ann. Math.* **47**, 731. Schrödinger, E. (1947) *Proc. R. Irish. Acad.* **31**, 163 and 205.

‡ Wheeler, J. A. and Misner, C. W. (1957) *Ann. Phys.* **2**, 525.

§ Heisenberg, W. *et al.* (1959) *Z. f. Naturforsch.* **14a**, 441.

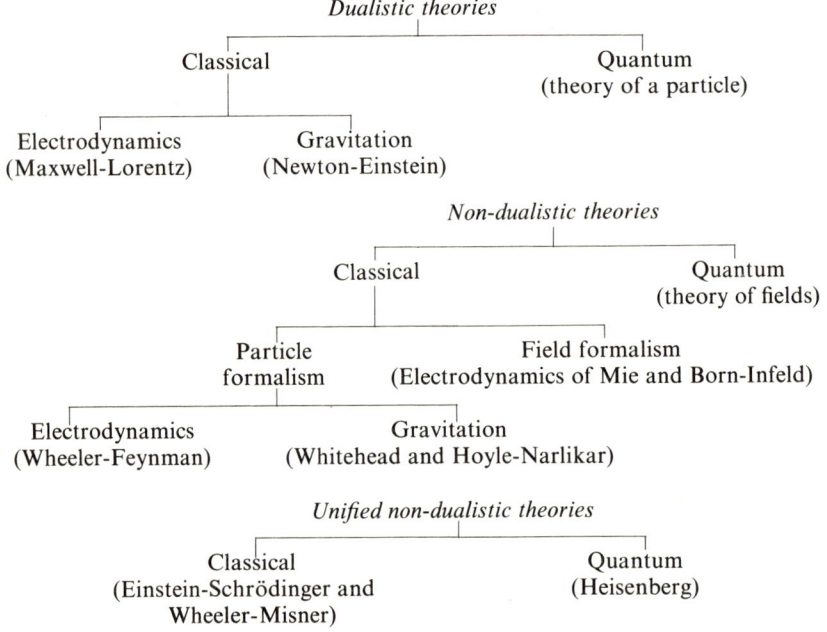

Part 1

Dualistic Theories

A. Classical Theories
B. Quantum Theory

A. Classical Theories

1. Electrodynamics

Space-time continuum. The physical phenomena (in this case electromagnetism) are assumed to take place in the background of a Minkowski space-time.

Let the co-ordinates of a space-time point be denoted by x^μ ($\mu = 1, 2, 3, 4$). Here x^i ($i = 1, 2, 3$) denotes the three space components and $x^4 = ict$, the imaginary time component.†

Our theory is assumed to be invariant under the homogeneous Lorentz group, the group of linear homogeneous transformations of the co-ordinates:

$$x^{\mu'} = L^{\mu'}_{\lambda} x^\lambda \tag{1}$$

or

$$x' = Lx \quad \text{(in matrix form)}$$

such that

$$L^{-1} = L^T, \quad \text{transpose of } L$$

Under a Lorentz transformation, a multicomponent tensorial field $\phi_A(x)$ transforms as

$$\phi_{A'}(Lx) = S^B_{A'} \phi_B(x) \tag{2}$$

where $S^B_{A'}$ is a linear function of L and the subscript A' (or B) stands for the general label for the components.

The metric interval ds will be given by

$$ds^2 = g_{\mu\lambda}\, dx^\mu\, dx^\lambda \tag{3}$$

where

$$g_{\mu\lambda} = \delta_{\mu\lambda}$$

so that there is no difference between the covariant and contravariant components of the same tensorial field: $\phi_A = \phi^A$.

Free particle. A free particle is characterized by its rest mass m and its position 4-vector $\xi_\mu = (\boldsymbol{\xi}, ic\tau)$, whose components depend on its eigentime l (in a rest system $l = \tau$). Let $\dot{\xi}_\mu = d\xi_\mu/dl$.

† Here c = velocity of light. We shall use the summation convention throughout; that is, repeated indices are to be summed. Real co-ordinates are also used frequently. The metric and its signature are explicitly stated where necessary.

The equation of motion of a free particle is assumed to follow from the action principle

$$\delta S_p = 0 \tag{4}$$

$$S_p = -mc \int \sqrt{[-(\dot{\xi}_\mu \dot{\xi}_\mu)]}\, dl$$

where S_p is an integral over a world line between any two space-time points; and ξ_μ is to be varied in such a way that $\delta\xi_\mu = 0$ at the end points. Equation (4) is equivalent to $\ddot{\xi}_\mu = 0$. Since

$$-\dot{\xi}_\mu \dot{\xi}_\mu = c\left(\frac{d\tau}{dl}\right)^2 \left(1 - \frac{v^2}{c^2}\right) \tag{5}$$

where $\mathbf{v} = d\boldsymbol{\xi}/d\tau$, we can also write

$$S_p = \int L_p \, d\tau \tag{6}$$

where $$L_p = -mc^2 \sqrt{1 - \left(\frac{v^2}{c^2}\right)}$$

the relativistic Lagrangian of a free particle. The momentum 3-vector is defined by

$$\mathbf{p} = \frac{\partial L_p}{\partial \mathbf{v}} = \frac{m\mathbf{v}}{\sqrt{1 - \left(\frac{v^2}{c^2}\right)}} \tag{7}$$

and the energy by

$$E = \mathbf{p} \cdot \mathbf{v} - L_p = \frac{mc^2}{\sqrt{1 - \left(\frac{v^2}{c^2}\right)}} \tag{8}$$

Thus E and p are related by

$$E^2/c^2 = p^2 + m^2c^2 \tag{9}$$

The electromagnetic field. To describe the motion of charged particles we are no longer able to say that one particle acts on another by means of instantaneous action-at-a-distance, because the forces acting on a particle at a given moment are not determined by positions alone of other particles at that moment. We say instead that a charged particle creates a field around itself and that the field exerts force on every other particle located in this field. The field thus acquires physical reality as an entity by itself.

Our *basic* assumption about the electromagnetic field is that it is characterized by an antisymmetric tensor-field $f_{\mu\lambda}(x) = -f_{\lambda\mu}(x)$ and that it satisfies the following equation *under all circumstances*:

$$f_{\mu\lambda,\nu}+f_{\lambda\nu,\mu}+f_{\nu\mu,\lambda} = 0 \tag{10}$$

Here $f_{\mu\lambda,\nu} = \partial f_{\mu\lambda}/\partial x^{\nu}$. The electric and magnetic fields $\mathbf{E}$ and $\mathbf{H}$ are then given by

$$\begin{aligned} \mathbf{H} &\equiv (f_{23}, f_{31}, f_{12}) \\ \mathbf{E} &\equiv \frac{1}{i}(f_{41}, f_{42}, f_{43}) \end{aligned} \tag{11}$$

Equation (10) implies that $f_{\mu\lambda}$ is derivable from a vector field A_{μ}, called the electromagnetic potential, according as

$$f_{\mu\lambda} = A_{\mu,\lambda}-A_{\lambda,\mu} \tag{12}$$

However, (12) does not define A_{μ} uniquely but only up to a *gauge transformation*:

$$A_{\mu} \rightarrow A'_{\mu} = A_{\mu}+\phi,_{\mu} \tag{13}$$

where ϕ is an arbitrary scalar field.

Free field. We consider the free-field situation, that is, electromagnetic field without any charged particles.

The free-field equations are to follow from the action principle

$$\left.\begin{aligned} \delta S_f &= 0 \\ S_f &= -\frac{1}{16\pi}\int f_{\mu\lambda}f_{\mu\lambda}\,dx \end{aligned}\right\} \tag{14}$$

Here $dx = d\mathbf{r}\,dt = dx_1\,dx_2\,dx_3\,dt$; the integral is over a closed volume and A_{μ} is to be varied in such a way that $\delta A_{\mu} = 0$ at the boundary. The field-equations are then

$$f_{\mu\lambda,\lambda} = 0 \tag{15}$$

Equations (15) and (10) characterize a free electromagnetic field.

Field-particle system. We now consider a field-particle system. The action function of such a system now consists of three terms:

$$S = S_p+S_f+S_{pf} \tag{16}$$

S_p and S_f defined above are functionals of particle co-ordinates and field potentials, respectively. S_{pf}, which describes the interaction between the field and particle, is a functional of both the field and particle variables:

$$S_{pf} = (e/c)\int dl\int \rho(x-\xi)[\dot{\xi}_{\mu}(A_{\mu}+A_{\mu}^{\text{ex}})]\,dx \tag{17}$$

Here the charge e characterizes the strength of interaction; ρ is a delta-like density function defined as follows:

$$\rho(x-\xi) = \delta(\mathbf{r}-\boldsymbol{\xi})\,\delta(t-\tau) \tag{18}$$

A_μ is the field generated by the particle and A_μ^{ex} is a *free external field* in which the charged particle is placed. As we are not interested in the field-equations of the external field (it satisfies (15)) we *do not* include $f_{\mu\lambda}^{\text{ex}}$ in S_f. The action principle now states

$$\delta S = 0 \tag{19}$$

The variation of ξ_μ yields the equation of motion of the particle under the action of the field and the variation of A_μ gives the field-equations of such a field-particle system.

We consider first the variation of ξ_μ. S_f, being independent of ξ_μ, does not contribute anything to (19) and we can write

$$\delta S' = 0, \qquad S' = \int L'\, dl \tag{20}$$

$$L' = -mc\sqrt{(-\dot{\xi}_\mu\dot{\xi}_\mu)}+(e/c)\int \rho(x-\xi)[\dot{\xi}_\mu(A_\mu+A_\mu^{\text{ex}})]\, dx$$

The Euler-Lagrange equation of the problem is

$$\frac{\delta L'}{\delta \xi_\mu} \equiv \frac{\partial L'}{\partial \xi_\mu}-\frac{d}{dl}\left(\frac{\partial L'}{\partial \dot{\xi}_\mu}\right) = 0 \tag{21}$$

According to (20) and (21) we have

$$\frac{mc\ddot{\xi}_\mu}{\sqrt{(-\dot{\xi}_\mu\dot{\xi}_\mu)}}+\frac{mc\dot{\xi}_\mu(\ddot{\xi}_\nu\dot{\xi}_\nu)}{(-\dot{\xi}_\lambda\dot{\xi}_\lambda)^{3/2}} = F_\mu+F_\mu^{\text{ex}} \tag{22}$$

where F_μ is the so-called self-force

$$F_\mu = (e/c)\dot{\xi}_\nu \int \rho(\xi-x)f_{\mu\nu}(x)\, dx = (e/c)\dot{\xi}_\nu f_{\mu\nu}(\xi) \tag{23}$$

and F_μ^{ex} is the Lorentz force due to the free external field.

$$F_\mu^{\text{ex}} = (e/c)\dot{\xi}_\nu f_{\mu\nu}^{\text{ex}}(\xi) \tag{24}$$

Because of antisymmetry of $f_{\mu\nu}$ and $f_{\mu\lambda}^{\text{ex}}$ we have $\dot{\xi}_\mu\dot{\xi}_\lambda f_{\mu\lambda} = \dot{\xi}_\mu\dot{\xi}_\lambda f_{\mu\lambda}^{\text{ex}} = 0$. This implies that $\dot{\xi}_\mu F_\mu = \dot{\xi}_\mu F_\mu^{\text{ex}} = 0$. Equation (22) implies then $\ddot{\xi}_\mu\dot{\xi}_\mu = 0$ or $\dot{\xi}_\mu\dot{\xi}_\mu = \text{const}$. The constant turns out to be $-c^2$ because for a rest particle $\dot{\xi}_k = 0$ and $\dot{t} = 1$. We have finally then the equation of motion

$$m\ddot{\xi}_\mu = F_\mu+F_\mu^{\text{ex}} \tag{25}$$

We now consider the field equations due to a charged particle. S_p does not contribute to the variation of A_μ, so that we can write

$$\delta S'' = 0, \qquad S'' = \int L''\, dx \tag{26}$$

$$L'' = -\frac{1}{16\pi} f_{\mu\lambda} f_{\mu\lambda} + \left(\frac{e}{c}\right) \int \rho(x-\xi) \dot{\xi}_\mu A_\mu \, dl$$

The Euler-Lagrange equation

$$\frac{\delta L''}{\delta A_\mu} \equiv \frac{\partial L''}{\partial A_\mu} - \frac{\partial}{\partial x^\nu}\left(\frac{\partial L}{\partial A_{\mu,\nu}}\right) = 0$$

is then the Maxwell's equation:

$$f_{\mu\lambda,\lambda} = \frac{4\pi e}{c} \int \rho(x-\xi) \dot{\xi}_\mu \, dl \tag{27}$$

The expression on the right is denoted by $(4\pi/c) j_\mu$ (j_μ is called the current 4-vector). Equation (27) becomes

$$f_{\mu\lambda,\lambda} = (4\pi/c) j_\mu \tag{28}$$

The antisymmetry of $f_{\mu\lambda}$ implies the continuity equation: $j_{\mu,\mu} = 0$.

Equations (10), (25) and (27) characterize completely a field-particle system in electrodynamics.

The freedom of gauge transformations (13) enables us to choose the following gauge condition, called the Lorentz gauge:

$$A_{\mu,\mu} = 0 \tag{29}$$

From (12), (27) and (29) we get

$$\square A_\mu = -\frac{4\pi e}{c} \int \rho(x-\xi) \dot{\xi}_\mu \, dl \tag{30}$$

The retarded solution of (30) is

$$A_\mu = e \int \dot{\xi}_\mu \frac{\delta(\tau - t + R'/c)}{cR'} \, dl \tag{31}$$

with $$\mathbf{R}' = \mathbf{r}(t) - \boldsymbol{\xi}(\tau)$$

Energy-momentum tensor of the field. From (25) and (27) we have

$$\int F_\mu \, dl = \frac{1}{4\pi} \int f_{\mu\nu} f_{\nu\lambda,\lambda} \, dx \tag{32}$$

With the help of (10) we can convert the right-hand integral into a divergence

$$\int F_\mu \, dl = \int T_{\mu\lambda,\,\lambda} \, dx \tag{33}$$

where
$$T_{\nu\mu} = T_{\mu\nu} = \frac{1}{4\pi} f_{\mu\lambda} f_{\lambda\nu} + \frac{1}{16\pi} \delta_{\mu\nu} f_{\alpha\beta} f_{\alpha\beta} \tag{34}$$

$T_{\mu\nu}$ is called the *energy-momentum tensor of the field.* Written in terms of the electric and magnetic fields we have for the various components:

$$T_{ik} = \frac{1}{4\pi} \left[E_i E_k + H_i H_k - \tfrac{1}{2}\delta_{ik}(H^2 + E^2) \right]; \qquad i, k = 1, 2, 3$$

$$(T_{14}, T_{24}, T_{34}) = -\frac{i}{4\pi} [\mathbf{E} \times \mathbf{H}], \qquad T_{44} = \frac{1}{8\pi} (E^2 + H^2) \tag{35}$$

It follows from (33) that for a free field (i.e. in the absence of any charges) $T_{\mu\lambda,\,\lambda} = 0$. Consequently

$$G_\mu = -\frac{1}{ic} \int T_{\mu 4} \, d\mathbf{r}$$

is a 4-vector, called the *field-momentum* 4-vector.

In a field-particle system we have from (25) and (33)

$$\int d[m\dot{\xi}_\mu + G_\mu] = \int F_\mu^{\text{ex}} \, dl + \int dt \oint T_{\mu k} \, ds_k \tag{36}$$

where the second term on the right-hand side contains a surface integral. In this case G_μ *is no longer a 4-vector in general.* This is one of the reasons for the failure of so-called electromagnetic mass theory of a charged particle, according to which a charged particle, e.g. an electron, has no inertial mass and its mass is entirely due to the inertia of its self-field. The success of such a theory must depend on the validity of the following conditions: (a) the self-field must have a finite energy so as to correspond to the finite masses of observed particles; (b) the self-field momentum must have the proper transformation properties of a particle momentum.

We have noted that the condition (b) is not satisfied in general and shall see later that the self-energy of a point particle is infinite.

Self-force and self-energy. One can compute directly the self-force which the field produced by the charge exerts on the charge itself. We first assume a certain charge distribution $\rho(\mathbf{r})$ and consider the non-relativistic motion ($v \ll c$) of such a particle along the x-axis. The equation of motion is then

$$m\ddot{x} = eE^{\text{ex}} + e \int \bar{E} \rho(\mathbf{r}) \, d\mathbf{r} \tag{37}$$

where E^{ex} is the external electric field and $\bar{E}$ the self-field which the charged particle itself creates.

$$\bar{E} = \int E\rho(\mathbf{r}')\,d\mathbf{r}'$$

with

$$\int \rho(\mathbf{r})\,d\mathbf{r} = 1 \tag{38}$$

To compute E we have from (31), $\left(\varphi = \frac{A_4}{i}, \mathbf{A} = (A_1, A_2, A_3)\right)$

$$\begin{aligned} E &= -\frac{\partial\varphi}{\partial x}-\frac{1}{c}\frac{\partial A_x}{\partial t} \\ \varphi &= e\int \frac{\delta(t'-t+R'/c)}{R'}\,dt' \\ A_x &= \frac{e}{c}\int v(t')\,\frac{\delta(t'-t+R'/c)}{R'}\,dt' \\ \mathbf{R}' &= \mathbf{r}-\mathbf{r}'(t') \end{aligned} \tag{39}$$

We expand the δ-function in a formal power series

$$\delta\left(t'-t+\frac{R'}{c}\right) = \delta(t'-t)+\frac{R'}{c}\dot{\delta}(t'-t)+\left(\frac{R'^2}{2c^2}\right)\ddot{\delta}(t'-t)+\frac{R'^3}{6c^3}\dddot{\delta}(t'-t)+\cdots \tag{40}$$

and get from (39)

$$\begin{aligned} E &= e\frac{R_x}{R^3}-\frac{e}{2c^2R}\left(\frac{R_x^2}{R^2}\dot{v}+\dot{v}\right)+\frac{2}{3}\frac{e}{c^3}\ddot{v}+\cdots \\ \mathbf{R} &= \mathbf{r}-\mathbf{r}'(t) \end{aligned}$$

so that finally the equation of motion becomes

$$m\ddot{x} = eE^{ex}-m_{el}\ddot{x}+\frac{2}{3}\frac{e^2}{c^3}\dddot{x}$$

where

$$m_{el} = \frac{2}{3}\frac{e^2}{c^2}\int d\mathbf{r}'\int \frac{\rho(\mathbf{r})\rho(\mathbf{r}')}{R}\,d\mathbf{r} \tag{41}$$

m_{el}, called the electromagnetic mass of the charged particle, can also be written as

$$m_{el} = \frac{4}{3}\frac{U_0}{c^2}$$

where U_0 is the electrostatic *self-energy* of the charge. For a point particle of course $U_0 \to \infty$ so that $m_{el} \to \infty$. On the other hand, if we assume that the charge is distributed on a spherical surface of radius r_0, $U_0 = e^2/2r_0$, so that

$$m_{el} = \frac{2}{3}\frac{e^2}{r_0 c^2}$$

r_0 is called the classical radius of electron. Of course, the choice of a surface distribution is quite arbitrary. If we ignore the structure of the particle and put $m+m_{el} = M$, the observed mass, we can provisionally neglect all effects of the self-force. But then, for a free particle we shall have (keeping only the $\dddot{x}$ term, which by the way is the only term in (41) independent of the structure of the particle)

$$-\frac{2}{3}\frac{e^2}{c^3}\dddot{x}+M\ddot{x} = 0$$

or

$$\ddot{x}(t) = \ddot{x}(0)\exp\left[\frac{3}{2}\frac{Mc^3}{e^2}\right]t$$

The motion of a free particle is thus self-accelerated.

We thus see that the Maxwell-Lorentz theory is beset with fundamental difficulties, the origin of which may be traced to the field-particle duality. We shall see later how these problems can be tackled in the framework of non-dualistic theories.

2. *Gravitation*

Principle of equivalence. Newton's theory of gravitation, being a pure particle theory, is based on instantaneous action-at-a-distance. Whereas the Maxwell-Lorentz theory of electromagnetism happened to be from the very outset Lorentz-covariant, the laws of gravitation were not. Attempts were made, notably by Poincaré and Minkowski, to make Newton's action-at-a-distance theory Lorentz-invariant. Since a field theory is more amenable to covariance, attention was next drawn to the problem of modifying the field-particle formalism of the Newtonian theory of gravitation, which consisted of (characteristic of all classical dualistic theories) (i) the field equations for the gravitational potential Φ (due to a distribution of matter of density ρ), the so-called Poisson's equation:

$$\nabla^2\Phi = 4\pi G\rho \quad (G = \text{const. of gravitation}) \tag{42}$$

and (ii) the equation of motion:

$$\frac{d^2\mathbf{r}}{dt^2} = -\text{grad}\,\Phi$$

Einstein postulated that the laws of gravitation be covariant under *general transformation of co-ordinates*. The motivation behind the postulate of general covariance is the so-called *principle of equivalence*. In the Newtonian theory a co-ordinate system in a *uniform* gravitational field is completely *equivalent* to a uniformly accelerated co-ordinate system *without any gravitational field*. In other words a non-inertial co-ordinate system is equivalent to a certain gravitational field. This equivalence is a consequence of the equality of inertial and gravitational mass of all particles. For *non-uniform* gravitational fields the principle of equivalence can be formulated as follows: *In any infinitely small space-time region there always exists a co-ordinate system in which gravitation has no influence on the motion of particles.*

According to Einstein both the postulate of general covariance and the principle of equivalence are best incorporated in the following mathematical scheme:

(a) *As far as gravitational phenomenon is concerned the physical space-time is a Riemannian manifold* V_4.
(b) *The metrical structure of* V_4 *is determined by the matter present in it;* in particular, the metric tensor $g_{\mu\lambda}$ of V_4 represents the gravitational potential. This constitutes in effect a *geometrization* of gravitation.

General covariance is then automatically assured in a V_4 and the principle of equivalence follows from the fact that in a V_4 there always exists a co-ordinate system such that $g_{\mu\lambda,\,\alpha} = 0$ at any given point.

The Riemannian space-time. A Riemannian manifold is a general manifold† with a certain metrical and affine structure. In a general manifold geometrical objects are classified according to their transformation properties under arbitrary co-ordinate transformations:

$$x^{\mu} \to x^{\mu'} = x^{\mu'}(x^{\mu})$$

with
$$\left|\frac{\partial x^{\mu'}}{\partial x^{\mu}}\right| \equiv |A^{\mu'}_{\mu}| \neq 0\ddagger \tag{43}$$

A multicomponent mixed tensor field then transforms as follows:

$$\xi^{\lambda'_1 \cdots \lambda'_p}_{\mu'_1 \cdots \mu'_q} = A^{\lambda'_1}_{\lambda_1} \cdots A^{\lambda'_p}_{\lambda_p} \cdot A^{\mu_1}_{\mu'_1} \cdots A^{\mu_q}_{\mu'_q} \xi^{\lambda_1 \cdots \lambda_p}_{\mu_1 \cdots \mu_q} \tag{44}$$

A mixed tensor density field on the other hand transforms as:

$$\mathscr{A}^{\lambda'_1 \cdots \lambda'_p}_{\mu'_1 \cdots \mu'_q} = |A^{\lambda}_{\lambda'}| A^{\lambda'_1}_{\lambda_1} \cdots A^{\lambda'_p}_{\lambda_p} A^{\mu_1}_{\mu'_1} \cdots A^{\mu_q}_{\mu'_q} \mathscr{A}^{\lambda_1 \cdots \lambda_p}_{\mu_1 \cdots \mu_q} \tag{45}$$

Consider now a contravariant vector field $\xi^{\mu}(x^{\lambda})$. In view of arbitrary

† By a general manifold we mean, more precisely, a *differentiable* manifold without any further structure; so that the manifold can be covered by co-ordinate neighbourhoods.

‡ $|A^{\mu'}_{\mu}| \equiv$ Jacobian of the transformation.

co-ordinate transformations (43) it is now impossible to compare the two values of the field ξ^μ and $\xi^\mu + d\xi^\mu$ at two infinitesimally neighbouring points x^λ and $x^\lambda + dx^\lambda$ in an invariant manner. However, it is possible to compare at the point $x^\lambda + dx^\lambda$ the value $\xi^\mu + d\xi^\mu$ with the so-called parallel-displaced vector $\xi^\mu + \delta\xi^\mu$ if $\nabla\xi^\mu \equiv d\xi^\mu - \delta\xi^\mu$ is assumed to be a vector. The general manifold acquires an affine structure if one assumes that the parallel displacement $\delta\xi^\mu$ is given by

$$\delta\xi^\mu = -\Gamma^\mu_{\alpha\beta}\xi^\alpha\, dx^\beta$$

with

$$\Gamma^\mu_{\alpha\beta} = \Gamma^\mu_{\beta\alpha} \tag{46}$$

The functions $\Gamma^\mu_{\alpha\beta}(x^\lambda)$ then characterize the affine structure. $\nabla\xi^\mu$ is called the covariant differential determined by the affine structure and

$$\xi^\mu_{;\lambda} \equiv \nabla_\lambda \xi^\mu = \xi^\mu_{,\lambda} + \Gamma^\mu_{\lambda\alpha}\xi^\alpha \tag{47}$$

is called the covariant derivative of ξ^μ. The transformation laws for $\Gamma^\mu_{\alpha\beta}$ are determined by the condition that $\xi^\mu_{;\lambda}$ be a tensor.

An affine structure defines a "path" $\xi^\mu(s)$ by the condition that its tangent vector $d\xi^\mu/ds$ be parallelly transferred to itself, i.e.

$$\delta\xi^\mu = d\xi^\mu$$

or

$$\frac{d^2\xi^\mu}{ds^2} + \Gamma^\mu_{\alpha\beta}\frac{d\xi^\alpha}{ds}\frac{d\xi^\beta}{ds} = 0 \tag{48}$$

Consider a point P whose co-ordinates in the unprimed system are $x^\lambda_{(P)}$. If we make the following transformation of co-ordinates:

$$x^{\lambda'} = x^\lambda - x^\lambda_{(P)} + \tfrac{1}{2}(\Gamma^\lambda_{\alpha\beta})_{(P)}(x^\alpha - x^\alpha_{(P)})(x^\beta - x^\beta_{(P)}) \tag{49}$$

one can show that $(\Gamma^{\mu'}_{\nu'\lambda'})_{(P)} = 0$. Thus the functions $\Gamma^\mu_{\nu\lambda}$ can be made to vanish at *any given point* by going over to the so-called geodesic co-ordinate system $x^{\lambda'}$ relative to the point according to (49).

Consider an infinitesimal parallelogram ABCD such that the co-ordinate differences from B to D and A to C are $\underset{1}{dx^\mu}$ and from A to B and C to D are $\underset{2}{dx^\mu}$, respectively. If we now parallelly displace a vector ξ^μ from A to D first *via* B and then *via* C, the difference of the two displaced vectors is given by

$$\Delta\xi^\mu = R^\mu_{\nu\alpha\beta}\xi^\nu \underset{1}{dx^\alpha}\underset{2}{dx^\beta}$$

where

$$R^\mu_{\nu\alpha\beta} = \Gamma^\mu_{\nu\beta,\alpha} - \Gamma^\mu_{\nu\alpha,\beta} + \Gamma^\mu_{\lambda\alpha}\Gamma^\lambda_{\nu\beta} - \Gamma^\mu_{\lambda\beta}\Gamma^\lambda_{\nu\alpha} \tag{50}$$

is called the curvature tensor. $R^\mu_{\nu\alpha\beta}$ has the following symmetry properties

$$\left.\begin{aligned} R^\mu_{\nu\alpha\beta} + R^\mu_{\nu\beta\alpha} &= 0 \\ R^\mu_{\nu\alpha\beta} + R^\mu_{\alpha\beta\nu} + R^\mu_{\beta\nu\alpha} &= 0 \end{aligned}\right\} \tag{51}$$

By choosing a geodesic co-ordinate system one can show that the curvature tensor satisfies the following Bianchi identity

$$R^{\mu}_{\nu\alpha\beta;\,\lambda}+R^{\mu}_{\nu\beta\lambda;\,\alpha}+R^{\mu}_{\nu\lambda\alpha;\,\beta} = 0 \tag{52}$$

We state an important theorem without proof.

Theorem. A necessary and sufficient condition for the existence of a co-ordinate system in which $\Gamma^{\mu}_{\alpha\beta} = 0$ *everywhere*† is $R^{\mu}_{\nu\alpha\beta} = 0$. In such a case we say that the affine structure is *integrable*.

We now introduce a metrical structure in the manifold by means of the metric tensor $g_{\mu\lambda} = g_{\lambda\mu}$. The *measure* of length of a vector ξ^{μ} is given by

$$\xi = g_{\mu\lambda}\xi^{\mu}\xi^{\lambda} \tag{53}$$

The measure of length of the infinitesimal element dx^{μ} is the "metric":

$$ds^2 = g_{\mu\lambda}\, dx^{\mu}\, dx^{\lambda} \tag{54}$$

$g_{\mu\lambda}$ is assumed to have an inverse so that one can define $g^{\mu\lambda}$ by $g_{\mu\nu}g^{\mu\lambda} = \delta^{\lambda}_{\nu}$. We raise and lower the tensor indices by $g_{\mu\lambda}$ and $g^{\mu\lambda}$. A metrical structure defines a geodesic curve $x^{\mu}(s)$ between any two points by the extremal length

$$\delta\int_1^2 ds = 0 \tag{55}$$

The equation for a geodesic is then given by

$$\frac{d^2x^{\mu}}{ds^2}+\begin{Bmatrix}\mu\\ \alpha\beta\end{Bmatrix}\frac{dx^{\alpha}}{ds}\frac{dx^{\beta}}{ds} = 0 \tag{56}$$

where

$$\begin{Bmatrix}\mu\\ \alpha\beta\end{Bmatrix} = \tfrac{1}{2}g^{\mu\lambda}(g_{\alpha\lambda,\,\beta}+g_{\beta\lambda,\,\alpha}-g_{\alpha\beta,\,\lambda}) \tag{57}$$

are the Christoffel symbols of the second kind.

A Riemannian manifold V_4 is then characterized by the following condition: its metrical structure is determined uniquely by its affine structure by the condition that the measure of length of a vector field is parallelly transferred to itself, i.e.

$$\delta\xi = \delta(g_{\mu\lambda}\xi^{\mu}\xi^{\lambda}) = 0 \tag{58}$$

where $\delta g_{\mu\lambda} = dg_{\mu\lambda}$ and $\delta\xi^{\mu}$ is given by (46). (58) then implies that

$$g_{\mu\lambda;\,\alpha} = 0 \tag{59}$$

which in turn implies that

$$\Gamma^{\mu}_{\alpha\beta} = \begin{Bmatrix}\mu\\ \alpha\beta\end{Bmatrix} \tag{60}$$

Thus in a V_4, according to (48) and (56), a "path" is also a geodesic.

† In the co-ordinate neighbourhood.

The curvature tensor in a V_4 is then

$$R^{\mu}_{\nu\alpha\beta} = \left\{ \begin{matrix} \mu \\ \nu\beta \end{matrix} \right\}_{,\alpha} - \left\{ \begin{matrix} \mu \\ \nu\alpha \end{matrix} \right\}_{,\beta} + \left\{ \begin{matrix} \mu \\ \lambda\alpha \end{matrix} \right\} \left\{ \begin{matrix} \lambda \\ \nu\beta \end{matrix} \right\} - \left\{ \begin{matrix} \mu \\ \lambda\beta \end{matrix} \right\} \left\{ \begin{matrix} \lambda \\ \nu\alpha \end{matrix} \right\} \tag{61}$$

The contracted curvature tensor

$$R^{\mu}_{\nu\alpha\mu} = R_{\nu\alpha} \tag{62}$$

is called the Ricci-Einstein tensor. The curvature scalar is defined by

$$R = g^{\nu\alpha} R_{\nu\alpha} \tag{63}$$

From the Bianchi identity (52) we get

$$(R^{\lambda}_{\mu} - \tfrac{1}{2}\delta^{\lambda}_{\mu} R)_{;\lambda} = 0 \tag{64}$$

According to (59) and (49), in a V_4 there always exists a geodesic co-ordinate system such that $g_{\mu\lambda,\alpha} = 0$ at any given point.

A V_4 with an integrable affine structure is called a flat-space. According to the theorem mentioned above, therefore, in a flat space there exists co-ordinate systems in which $g_{\mu\lambda,\alpha} = 0$ *everywhere.*

The free gravitational field. As in the electromagnetic case, we consider first the free gravitational field in absence of matter. According to Einstein's principle of geometrization, the metric tensor $g_{\mu\lambda}$ represents the gravitational field (potential to be exact). The free-field equations are then to follow from a variational principle:

$$\delta \int \mathscr{W}\, dx = 0 \tag{65}$$

Here $\mathscr{W}$ is some scalar-density function of $g_{\mu\lambda}$ and their derivations $g_{\mu\lambda,\alpha}$ only. We can then write

$$\delta \int \mathscr{W}\, dx = \int \mathscr{W}^{\mu\lambda} \cdot \delta g_{\mu\lambda}\, dx = 0 \tag{66}$$

from which follow the Euler-Lagrange equations

$$\mathscr{W}^{\mu\lambda} \equiv \frac{\delta \mathscr{W}}{\delta g_{\mu\lambda}} \equiv \frac{\partial \mathscr{W}}{\partial g_{\mu\lambda}} - \frac{\partial}{\partial x^{\alpha}} \left(\frac{\partial \mathscr{W}}{\partial g_{\mu\lambda,\alpha}} \right) = 0 \tag{67}$$

In a variational principle of this type there always follow certain identities because of the invariance of the action integral under co-ordinate transformations. The invariance is guaranteed by the fact that $\mathscr{W}$ is a scalar-density. It suffices to consider only infinitesimal co-ordinate transformations:

$$x^{\mu'} = x^{\mu} + \delta x^{\mu}$$

Then

$$\delta g_{\mu\lambda} = -g_{\mu\alpha}(\delta x^{\alpha})_{,\lambda} + g_{\lambda\alpha}(\delta x^{\alpha})_{,\mu} + g_{\mu\lambda,\alpha}\, \delta x^{\alpha} \tag{68}$$

If we substitute (68) in (66) we get

$$\int [2\mathscr{W}^{\mu}_{\lambda}(\delta x^{\lambda})_{,\mu} + \mathscr{W}^{\mu\lambda} g_{\mu\lambda,\alpha}\,\delta x^{\alpha}]\,dx = 0$$

Since δx^{α} is arbitrary we obtain after a partial integration

$$\mathscr{W}^{\mu}_{\lambda,\mu} - \tfrac{1}{2}\mathscr{W}^{\nu\alpha} g_{\nu\alpha,\lambda} = 0$$

which can be written as

$$\mathscr{W}^{\mu}_{\lambda;\mu} \equiv \mathscr{W}^{\mu}_{\lambda,\mu} - \mathscr{W}^{\beta}_{\alpha}\begin{Bmatrix}\alpha\\ \lambda\beta\end{Bmatrix} = 0 \tag{69}$$

For a free gravitational field we assume that

$$\mathscr{W} = R\sqrt{-g} = g^{\mu\lambda}R_{\mu\lambda}\sqrt{-g} \tag{70}$$

The factor $\sqrt{-g}$ makes $\mathscr{W}$ a scalar-density. The minus sign under the square root is due to the fact that $g = |g_{\mu\lambda}|$ is negative for real space-time co-ordinates. From (70) and (67) the free-field equations are then

$$\mathscr{W}^{\mu\lambda} \equiv -\sqrt{(-g)}\,(R^{\mu\lambda} - \tfrac{1}{2}g^{\mu\lambda}R) = 0 \tag{71}$$

which can also be written as

$$R_{\mu\lambda} = 0 \tag{72}$$

The identities (69) for this choice of $\mathscr{W}$ turn out to be identical with (64).

Equation (72) is to be considered as the relativistic generalization of the Newtonian free-field equations, e.g. $\nabla^2\Phi = 0$.

Field-particle system. Our problem is now to find the relativistic generalization of equation (42). We consider an incoherent distribution of matter of rest-mass density ρ. The energy-momentum tensor of matter is given by

$$T^{\mu\lambda} = \rho u^{\mu} u^{\lambda} \tag{73}$$

where $u^{\lambda} \equiv u^{\lambda}(x)$ is the velocity 4-vector dx^{λ}/ds at the point x. The field equation of such a system is then

$$\mathscr{W}^{\mu\lambda} = \kappa\mathscr{T}^{\mu\lambda}; \qquad \mathscr{T}^{\mu\lambda} = \sqrt{(-g)}\,T^{\mu\lambda} \tag{74}$$

or

$$R_{\mu\lambda} - \tfrac{1}{2}g_{\mu\lambda}R = -\kappa T_{\mu\lambda}$$

where κ is a constant. Equation (74) is to be considered as the relativistic generalization of Poisson's equation. It is also the gravitational analogue of the Maxwell's equations.

On the other hand the equation of motion of a test particle in a given background gravitational field is given by the geodesic equation (56)

$$\frac{d^2x^{\mu}}{ds^2} + \begin{Bmatrix}\mu\\ \alpha\beta\end{Bmatrix}\frac{dx^{\alpha}}{ds}\frac{dx^{\beta}}{ds} = 0$$

We now show that equations (56) and (74) lead us, as a limiting case, to the Newtonian equations (42). Consider first the equation of motion (56) of a test-particle under the following conditions: (i) its velocity is small compared to that of light; (ii) the background gravitational field is weak. This means $g_{\mu\lambda}$ differs only slightly from their Galilean values (Latin indices take the values 1, 2, 3, whereas Greek indices 0, 1, 2, 3);

$$\eta_{ik} = -1 \quad \text{for} \quad i, k = 1, 2, 3, \qquad \eta_{00} = +1$$

$$\eta_{\mu\lambda} = 0 \quad \text{for} \quad \mu \neq \lambda, \qquad x = (\mathbf{x}, ct)$$

so that we have from (56)

$$\frac{d^2x^i}{dt^2} = -c^2 \begin{Bmatrix} i \\ 00 \end{Bmatrix} \quad \text{for} \quad i = 1, 2, 3;$$

(iii) the field is quasi-static, so that $g_{\mu\lambda, 0}$ can be neglected. Then

$$\begin{Bmatrix} i \\ 00 \end{Bmatrix} \simeq \tfrac{1}{2} g_{00, i}$$

Under these conditions (56) reduces to the Newtonian equations of motion

$$\frac{d^2x^i}{dt^2} = -\frac{\partial \Phi}{\partial x^i}$$

where we have put

$$\Phi = \tfrac{1}{2}c^2(g_{00} - 1) \quad \text{or} \quad g_{00} = 1 + 2\Phi/c^2 \tag{75}$$

The additive constant in Φ is determined by the conditions that $\Phi \to 0$ as $g_{00} \to \eta_{00}$.

Consider now the field equations (74) under the same conditions. We have then

$$T_{\mu\lambda} = 0 \quad \text{except for} \quad T_{00} = \rho c^2$$

and

$$R_{00} = -\begin{Bmatrix} \alpha \\ 00 \end{Bmatrix}_{,\alpha} \simeq -\tfrac{1}{2}\nabla^2 g_{00} = -\frac{\nabla^2 \Phi}{c^2}$$

so that (74) reduces to the Poisson's equation

$$\nabla^2 \Phi = \tfrac{1}{2}\kappa c^4 \rho \tag{76}$$

Comparing (76) with (42) we find that

$$\kappa = \frac{8\pi G}{c^4}$$

The exact solutions of (74) and (56) then indicate the deviations from the Newtonian theory. These deviations have been experimentally confirmed, for example, in the special case of the gravitational field due to a point-particle at rest. We mention the so-called Schwarzschild's solution of (72) for a static spherically symmetric field due to point-particle at the origin (in polar co-ordinates).

$$ds^2 = \left\{ \frac{dr^2}{1-\dfrac{2m}{r}} + r^2\, d\theta^2 + r^2 \sin^2\theta\, d\phi^2 \right\} + c^2\left(1-\frac{2m}{r}\right) dt^2 \tag{77}$$

where $m = G\dfrac{M}{c^2}$, M = mass of the particle

Energy-momentum pseudo-tensor of the field. From (69) and (74) we obtain the following identity for the matter tensor-density:

$$\mathscr{T}^{\mu}_{\lambda;\mu} \equiv \mathscr{T}^{\mu}_{\lambda,\mu} - \mathscr{T}^{\beta}_{\alpha}\begin{Bmatrix}\alpha\\ \lambda\beta\end{Bmatrix} = 0 \tag{78}$$

Equation (78) is to be considered *formally* as a conservation identity. Being a tensorial equation it is generally covariant; but it is precisely its general covariance which makes any physical interpretation impossible. Moreover, since we are dealing with a matter-field system we should expect a conservation law for the matter-tensor plus that of the gravitational field. It is, however, possible to convert the second term in (78) into an ordinary divergence. To see this we note from (67) and (70) that $\mathscr{W}^{\mu\lambda}$ is derivable from the action function $R\sqrt{-g}$ as its Hamiltonian derivative. Since the addition of an ordinary divergence to the action function does not change the action integral, $\mathscr{W}^{\mu\lambda}$ is *also* derivable from an equivalent action function $-\mathscr{L}$ defined as follows:

$$\begin{aligned} \mathscr{L} &= \sqrt{(-g)}\, g^{\mu\lambda}\mathscr{L}_{\mu\lambda} \\ \mathscr{L}_{\mu\lambda} &= \begin{Bmatrix}\beta\\ \alpha\lambda\end{Bmatrix}\begin{Bmatrix}\alpha\\ \mu\beta\end{Bmatrix} - \begin{Bmatrix}\alpha\\ \alpha\beta\end{Bmatrix}\begin{Bmatrix}\alpha\\ \mu\lambda\end{Bmatrix} \end{aligned} \tag{79}$$

The difference between $R\sqrt{-g}$ and $-\mathscr{L}$ is then an ordinary divergence. We have then from (67) and (74)

$$\mathscr{T}^{\mu\lambda} = \frac{\mathscr{W}^{\mu\lambda}}{\kappa} = -\frac{1}{\kappa}\frac{\delta\mathscr{L}}{\delta g_{\mu\lambda}} \equiv \frac{1}{\kappa}\left[\frac{\partial}{\partial x^{\alpha}}\left(\frac{\partial\mathscr{L}}{\partial g_{\mu\lambda,\alpha}}\right) - \frac{\partial\mathscr{L}}{\partial g_{\mu\lambda}}\right]$$

The second term in (78) can therefore be written as

$$\begin{aligned}
\tfrac{1}{2}\mathcal{T}^{\nu\alpha}g_{\nu\alpha,\lambda} &= \frac{1}{2\kappa}\left[\frac{\partial}{\partial x^{\beta}}\left(\frac{\partial\mathcal{L}}{\partial g_{\nu\alpha,\beta}}\right)g_{\nu\alpha,\lambda}-\frac{\partial\mathcal{L}}{\partial g_{\nu\alpha}}g_{\nu\alpha,\lambda}\right] \\
&= \frac{1}{2\kappa}\left[\frac{\partial}{\partial x^{\beta}}\left(\frac{\partial\mathcal{L}}{\partial g_{\nu\alpha,\lambda\beta}}g_{\nu\alpha,\lambda}\right)-\frac{\partial\mathcal{L}}{\partial g_{\nu\alpha,\beta}}g_{\nu\alpha,\beta,\lambda}-\frac{\partial\mathcal{L}}{\partial g_{\nu\alpha}}g_{\nu\alpha,\lambda}\right] \\
&= \frac{1}{2\kappa}\left[\frac{\partial}{\partial x^{\beta}}\left(\frac{\partial\mathcal{L}}{\partial g_{\nu\alpha,\beta}}g_{\nu\alpha,\lambda}\right)-\frac{\partial\mathcal{L}}{\partial x^{\lambda}}\right]
\end{aligned}$$

Therefore (78) can be written finally as

$$(\mathcal{T}^{\mu}_{\lambda}+t^{\mu}_{\lambda})_{,\mu} = 0 \tag{80}$$

where

$$t^{\mu}_{\lambda} = \frac{1}{2\kappa}\left(\delta^{\mu}_{\lambda}\mathcal{L}-\frac{\partial\mathcal{L}}{\partial g_{\nu\alpha,\mu}}g_{\nu\alpha,\lambda}\right) \tag{81}$$

We can thus regard t^{μ}_{λ} as the contribution of the gravitational field to the total energy-momentum of the matter-field system. t^{μ}_{λ} is, however, *not* a tensor-density (hence called pseudo-tensor). However, (80) holds in *every co-ordinate system* provided that the pseudo-tensor is defined in that system according to (81). Since t^{μ}_{λ} does not depend on derivatives of $g_{\mu\lambda}$ higher than the first, it can be made to vanish at a point in a geodesic co-ordinate system. Moreover, in the field of an isolated mass-point it can be made to vanish everywhere by a suitable transformation of co-ordinates. Einstein has shown that one cannot assign any physical meaning to the values of t^{μ}_{λ}, i.e. it is impossible to localize energy-momentum in a gravitational field in a generally covariant and physically satisfactory manner. The physical significance of (80) lies in the fact that it allows us to calculate the change in material energy of a closed system in a simple way.

Equations of motion. A characteristic feature of all classical dualistic theories is that it is always necessary to postulate two sets of equations: the field equations, which enable us to calculate the field due to a given source-distribution and the equations of motion of a test particle in a given background field.

Of the two theories we have discussed so far, the field equations of gravitation differ from that of electromagnetism in one fundamental respect; they are non-linear.

Consider, for example, a system of charged particles. One can in principle separate from the total electromagnetic field, the field produced by a single particle. Both the total electromagnetic field and the separate individual field of a single particle or their difference satisfies the Maxwell's equations which are linear. On the other hand, the Lorentz force acting on a charged particle is completely independent of the Maxwell's equation. Therefore the introduction of a non-electromagnetic force does not modify the field equations.

In the case of the gravitational field this separation is not possible because of the non-linearity of the field equations of gravitation. Moreover, the introduction of a non-gravitational force on a particle modifies the energy momentum tensor $T_{\mu\lambda}$ and consequently the field, according to (74).

In contrast to the Maxwell-Lorentz theory the non-linearity of the gravitational field equations enables us to deduce the equations of motion of a test particle from the field equations alone, so that the geodesic postulate (56) becomes superfluous. The non-linearity is, however, a *necessary but not sufficient* condition for this deduction.

A more difficult problem is to derive the equations of motion of *gravitating* particles (as opposed to test particles in a background field) in their combined gravitational field. This was solved by Einstein, Infeld and Hoffmann† on the assumption that the particles be represented as point singularities of the gravitational field equations in empty space (i.e. (74) without $T_{\mu\lambda}$). We consider first the simpler problem of motion of test particles in a given background field.

(a) *Equation of motion of test particles from the matter tensor.* We start from the equations (74) and (78).

$$R_{\mu\lambda}-\tfrac{1}{2}g_{\mu\lambda}R = -\kappa T_{\mu\lambda}$$

$$T^{\mu\lambda} = \rho u^{\mu}u^{\lambda}, \qquad T^{\mu\lambda}{}_{;\lambda} = 0 \tag{82}$$

We have then

$$(\rho u^{\mu}u^{\lambda})_{;\lambda} = (\rho u^{\lambda})_{;\lambda}u^{\mu}+(\rho u^{\lambda})u^{\mu}{}_{;\lambda} = 0 \tag{83}$$

Multiplying by u_{μ} we get

$$(\rho u^{\lambda})_{;\lambda}+(\rho u^{\lambda}u_{\mu})u^{\mu}{}_{;\lambda} = 0 \tag{84}$$

because

$$u_{\mu}u^{\mu} = 1$$

Moreover

$$(u_{\mu}u^{\mu})_{;\lambda} = 0 = 2u^{\mu}{}_{;\lambda}u_{\mu}$$

so that the second term in (84) vanishes and we are left with the relativistic generalization of the continuity equation

$$(\rho u^{\lambda})_{;\lambda} = 0 \tag{85}$$

Finally, from (83)

$$\rho u^{\lambda}u^{\mu}{}_{;\lambda} = 0 \tag{86}$$

† Einstein, A., Infeld, L. and Hoffmann, B. (1938) *Ann. Math.* **39**, 65; Einstein, A. and Infeld, L. (1949) *Can. J. Math.* **1**, 209.

If we have a single point test-particle of unit mass whose path is given by $\bar{x}^\mu = \bar{x}^\mu(s)$, we can write formally†

$$\rho(x) = \delta(x-\bar{x})/\sqrt{-g} \tag{87}$$

so that

$$\int \rho\sqrt{(-g)}\,dx = 1$$

(86) can be written as

$$\frac{\delta(x-\bar{x})}{\sqrt{-g}}\frac{d^2x^\mu}{ds^2}+\begin{Bmatrix}\mu\\ \alpha\beta\end{Bmatrix}\frac{dx^\alpha}{ds}\frac{dx^\beta}{ds} = 0$$

Multiplying by $\sqrt{-g}$ and integrating over the space-time we get the geodesic equation of motion

$$\frac{d^2\bar{x}^\mu}{ds^2}+\begin{Bmatrix}\mu\\ \alpha\beta\end{Bmatrix}_{\bar{x}}\frac{d\bar{x}^\alpha}{ds}\frac{d\bar{x}^\beta}{ds} = 0 \tag{88}$$

where $\begin{Bmatrix}\mu\\ \alpha\beta\end{Bmatrix}_{\bar{x}}$ is to be computed at $\bar{x}$.

(b) *Equations of motion of gravitating particles from the free-field equations.* Instead of using the field equations (74) where matter is represented by the energy-momentum tensor $T_{\mu\lambda}$, one considers the free-field equations in empty space and represents particles as point-singularities of the gravitational field.

Let

$$g_{\mu\nu} = \eta_{\mu\nu}+h_{\mu\nu}, \qquad g^{\mu\nu} = \eta^{\mu\nu}+h^{\mu\nu} \tag{89}$$

where $\eta_{\mu\nu}$ are the Galilean values of the metric

$$\eta_{mn} = -\delta_{mn}, \qquad \eta_{00} = 1, \qquad \eta_{0i} = 0$$

and

$$g_{\mu\sigma}g^{\mu\nu} = \delta^\nu_\sigma$$

It is convenient to introduce the following quantities in place of $h_{\mu\nu}$

$$\gamma_{\mu\nu} = h_{\mu\nu}-(1/2)\eta_{\mu\nu}\eta^{\sigma\rho}h_{\sigma\rho} \tag{90}$$

The equations of the gravitational field in empty space

$$R_{\mu\nu} = 0 \tag{91}$$

† Sen, D. K. (1961) *Nuovo Cim.* **21**, 184.

then become (summation is implied over *all* repeated indices)

$$\begin{aligned} \Phi_{00}+2\Lambda_{00} &= 0 \\ \Phi_{0i}+2\Lambda_{0i} &= 0 \\ \Phi_{ik}+2\Lambda_{ik} &= 0 \end{aligned} \tag{92}$$

where

$$\begin{aligned} \Phi_{00} &= -\gamma_{00,i,i} \\ \Phi_{0i} &= -\gamma_{0i,k,k}+\gamma_{0j,j,i} \\ \Phi_{ik} &= -\gamma_{ik,j,j}+\gamma_{ij,k,j}+\gamma_{kj,i,j}-\delta_{ik}\gamma_{jm,j,m} \end{aligned} \tag{93}$$

and

$$\begin{aligned} 2\Lambda_{00} &= \gamma_{sr,s,r}+2\Lambda'_{00} \\ 2\Lambda_{0m} &= \gamma_{ms,s,0}-\gamma_{00,m,0}+2\Lambda'_{0m} \\ 2\Lambda_{mn} &= -\gamma_{0m,0,n}-\gamma_{0n,0,m}+2\delta_{mn}\gamma_{0s,0,s} \\ &\quad +\gamma_{mn,0,0}-\delta_{mn}\gamma_{00,0,0}+2\Lambda'_{mn} \end{aligned} \tag{94}$$

The linear terms are written out explicitly and $\Lambda'_{\mu\lambda}$ stands for non-linear terms in $\gamma_{\alpha\beta}$.

Let us now assume that there are p particle sources of the field which we regard as singularities of the field. Suppose that the path of the ith singularity is given by

$$\overset{i}{\xi}_k(x^0), \qquad i = 1, 2, \ldots, p \tag{95}$$

The gravitational field depends on the field point x as well as on the ξ's and their time derivatives.

At an arbitrary moment x^0 we surround the ith singularity, and it alone, by a two-dimensional closed surface $\overset{i}{\Omega}$ and consider the surface integral:

$$\int_{\overset{i}{\Omega}} (\Phi_{\mu k}+2\Lambda_{\mu k})n_k \, d\Omega = 0 \tag{96}$$

From (93) it follows that $\Phi_{\mu k}$ can be written as

$$\Phi_{\mu k} = F_{(\mu)kl,l} \tag{97}$$

where

$$F_{(\mu)kl} = \gamma_{\mu l,k}-\gamma_{\mu k,l}-\delta_{\mu k}\gamma_{lr,r}+\delta_{\mu l}\gamma_{kr,r} \tag{98}$$

$$F_{(\mu)kl} = -F_{(\mu)lk} \tag{99}$$

Thus
$$\int_{\Omega}^{i} \Phi_{\mu k} n_k \, d\Omega \equiv \int_{\Omega}^{i} F_{(\mu)kl,\,l} n_k \, d\Omega \tag{100}$$

The above integral is independent of the shape of the surface $\overset{i}{\Omega}$ because

$$F_{(\mu)kl,\,l,\,k} = 0$$

We can also write it in the form

$$\int_{\Omega}^{i} \operatorname{curl}_n \mathbf{A} \, d\Omega \tag{101}$$

where $\qquad F_{(\mu)23} = A_1, \qquad F_{(\mu)31} = A_2, \qquad F_{(\mu)12} = A_3$

But (101) can be changed into a line integral over the rim of the surface, which vanishes as $\overset{i}{\Omega}$ is closed. Thus we deduce from (96) that

$$\int_{\Omega}^{i} \Lambda_{\mu k} n_k \, d\Omega = 0 \tag{102}$$

Furthermore, $\Phi_{\mu n,\, n} = 0$. Therefore also $\Lambda_{\mu n,\, n} = 0$, so that the above surface integral does not depend on the nature of the surface. The $4p$ relations (102) cannot therefore give us any relation between the co-ordinates of the field. They can only give us relations between the co-ordinates of the singularities and their time derivatives. These are then the required equations of motion of the singularities.

The above equations so far are exact and no approximations have been made yet on $h_{\mu\nu}$. The Einstein-Infeld-Hoffmann method consists of an approximation procedure based on the following expansion† of $\gamma_{\mu\nu}$ in power series of a small parameter λ.

$$\begin{aligned} \gamma_{00} &= \lambda^2 \underset{2}{\gamma_{00}} + \lambda^4 \underset{4}{\gamma_{00}} + \cdots \\ \gamma_{0m} &= \lambda^3 \underset{3}{\gamma_{0m}} + \lambda^5 \underset{3}{\gamma_{0m}} + \cdots \\ \gamma_{mn} &= \lambda^4 \underset{4}{\gamma_{mn}} + \lambda^6 \underset{6}{\gamma_{mn}} + \cdots \end{aligned} \tag{103}$$

The indices below indicate the order of the coefficients. Furthermore, the field is supposed to be *quasi-static*, that is, the derivatives with respect to x^0 are assumed to be of a higher order than the space derivatives. Let $\tau = x^0 \lambda$. We can treat the derivatives with respect to τ on the same footing as the space derivatives. We shall have

$$f(x)_{,0} \equiv \frac{\partial f}{\partial x^0} = \frac{\partial f}{\partial \tau} \lambda \equiv \lambda f_{|0} \tag{104}$$

† The choice of different powers in the expansion can be justified heuristically.

where the stroke denotes differentiation with respect to τ. Thus the comma can be replaced by the stroke differentiation if the power of λ is raised. To express this we use a suffix under the index zero. Thus

$$\lambda^{2l}\underset{2l}{\gamma}_{mn,\,0} = \lambda^{2l+1}\underset{2l}{\gamma}_{mn|\underset{1}{0}},;$$

$$\lambda^{2l}\underset{2l}{\gamma}_{mn,\,0,\,0} = \lambda^{2l+2}\underset{2l}{\lambda}_{mn|0|\underset{2}{0}}$$

From now on all differentiations are with respect to (τ, x^1, x^2, x^3) and they will be denoted by strokes. Thus

$$\gamma_{\mu\nu,\,k} \equiv \gamma_{\mu\nu|k}, \qquad \gamma_{\mu\nu,\,0} \equiv \lambda\gamma_{\mu\nu|\underset{1}{0}}$$

We now substitute the development (103) in (92) and get in the various orders of approximation:

$$\begin{aligned} \underset{2l-2}{\Phi_{00}} + 2\underset{2l-2}{\Lambda_{00}} &= 0 \\ \underset{2l-2}{\Phi_{0m}} + 2\underset{2l-1}{\Lambda_{0m}} &= 0 \\ \underset{2l}{\Phi_{mn}} + 2\underset{2l}{\Lambda_{mn}} &= 0 \end{aligned} \tag{105}$$

where

$$\begin{aligned} \underset{2l-2}{\Phi_{00}} &= -\underset{2l-2}{\gamma_{00|r|r}} \\ \underset{2l-1}{\Phi_{0m}} &= -\underset{2l-1}{\gamma_{0m|r|r}} + \underset{2l-1}{\gamma_{0r|m|r}} \\ \underset{2l}{\Phi_{mn}} &= -\underset{2l}{\gamma_{mn|r|r}} + \underset{2l}{\gamma_{mr|n|r}} + \underset{2l}{\gamma_{nr|m|r}} \\ &\quad -\delta_{mn}\underset{2l}{\gamma_{rs|r|s}} \end{aligned} \tag{106}$$

and

$$\begin{aligned} 2\underset{2l-2}{\Lambda_{00}} &= \underset{2l-2}{\gamma_{rs|r|s}} + 2\underset{2l-2}{\Lambda'_{00}} \\ 2\underset{2l-1}{\Lambda_{0m}} &= -\underset{2l-2}{\gamma}_{00|\underset{1}{0}|m} + \underset{2l-2}{\gamma}_{mr|\underset{1}{0}|r} + 2\underset{2l-1}{\Lambda'_{0m}} \\ 2\underset{2l}{\Lambda_{mn}} &= -\underset{2l-1}{\gamma}_{0m|\underset{1}{0}|n} - \underset{2l-1}{\gamma}_{0n|\underset{1}{0}|m} + 2\delta_{mn}\underset{2l-1}{\gamma}_{0r|\underset{1}{0}|r} \\ &\quad + \underset{2l-2}{\gamma}_{mn|0|\underset{2}{0}} - \delta_{mn}\underset{2l-2}{\gamma}_{00|0|\underset{2}{0}} + 2\underset{2l}{\Lambda'_{mn}} \end{aligned} \tag{107}$$

Suppose now that

$$\underset{2}{\gamma_{00}} \cdots\cdots \underset{2l-4}{\gamma_{00}};$$

$$\underset{3}{\gamma_{0m}} \cdots\cdots \underset{2l-3}{\gamma_{0m}};$$

$$\underset{4}{\gamma_{mn}} \cdots\cdots \underset{2l-2}{\gamma_{mn}}$$

are known; then equations (105) when solved give us

$$\underset{2l-2}{\gamma_{00}}, \quad \underset{2l-1}{\gamma_{0m}}, \quad \underset{2l}{\gamma_{mn}}$$

entirely in terms of the known quantities. Thus the field can be determined to any approximation.

It is possible to simplify the equations by adopting a suitable co-ordinate system.

If

$$\underset{2l-2}{\gamma^*_{00}}, \quad \underset{2l-1}{\gamma^*_{0m}}, \quad \underset{2l}{\gamma^*_{mn}}$$

are solutions of (105) then, it can be shown by straightforward substitution that any

$$\begin{aligned}
\underset{2l-2}{\gamma_{00}} &= \underset{2l-2}{\gamma^*_{00}} \\
\underset{2l-1}{\gamma_{0m}} &= \underset{2l-1}{\gamma^*_{0m}} + \underset{2l-1}{a_{0|m}} \\
\underset{2l}{\gamma_{mn}} &= \underset{2l}{\gamma^*_{mn}} + \underset{2l}{a_{m|n}} + \underset{2l}{a_{n|m}} - \delta_{mn}\underset{2l}{a_{r|r}} + \delta_{mn}\underset{2l-1\;\;1}{a_{0|0}}
\end{aligned} \tag{108}$$

with *arbitrary* $\underset{2l-1}{a_0}$, $\underset{2l}{a_m}$ are also solutions of (105). Thus at each approximation step one can impose four co-ordinate conditions on the field. The co-ordinate conditions we take are the following.

$$\begin{aligned}
\underset{2l-2\;1}{\gamma_{00|0}} - \underset{2l-1}{\gamma_{0s|s}} &= 0 \\
\underset{2l}{\gamma_{mn|n}} &= 0
\end{aligned} \tag{109}$$

(105) then becomes

$$\begin{aligned}
\underset{2l-2}{\gamma_{00|r|r}} &= 2\underset{2l-2}{\Lambda'_{00}} \\
\underset{2l-1}{\gamma_{0m|r|r}} &= 2\underset{2l-1}{\Lambda'_{0m}} \\
\underset{2l}{\gamma_{mn|r|r}} &= -\underset{2l-1\;1}{\gamma_{0m|0|n}} - \underset{2l-1\;1}{\gamma_{0n|0|0m}} + \delta_{mn}\underset{2l-2\;2}{\gamma_{00|0|0}} + \underset{2l-2\;2}{\gamma_{mn|0|0}} \\
&\quad + 2\underset{2l}{\Lambda'_{mn}} \\
&= 2\underset{2l}{\Lambda_{mn}}
\end{aligned} \tag{110}$$

and the surface integrals that must vanish and which give the equations of motion are

$$\int_{\overset{i}{\Omega}} (2\underset{2l-1}{\Lambda'_{0m}} - \underset{2l-2}{\gamma}{}_{00|\underset{1}{0}|m})n_m \, d\Omega = 0$$

$$\int_{\overset{i}{\Omega}} 2\underset{2l}{\Lambda_{nm}} n_m \, d\Omega = 0 \tag{111}$$

We now consider the Newtonian approximation with $l = 2$. Equation (110) becomes

$$\underset{2}{\gamma_{00|r|r}} = 0$$

$$\underset{3}{\gamma_{0m|r|r}} = 0 \tag{112}$$

$$\underset{4}{\gamma_{nm|r|r}} = 2\underset{4}{\Lambda_{nm}}$$

and the co-ordinate conditions are

$$\underset{3}{\gamma_{0r|r}} - \underset{2}{\gamma}{}_{00|\underset{1}{0}} = 0$$

$$\underset{4}{\gamma_{mr|r}} = 0 \tag{113}$$

The character of our solution depends essentially on the choice of the harmonic function we take as the solution of γ_{00} in (112). As we are interested in particle-like solutions we exclude all dipoles and poles of higher order and take

$$\underset{2}{\gamma_{00}} = 2\phi, \qquad \phi = \sum_{i=1}^{p} (-2\underset{2}{\overset{i}{m}}\overset{i}{\psi})$$

$$\overset{i}{\psi} = [(x_r - \overset{i}{\xi}_r)]^{-1/2} = (\overset{i}{r})^{-1} \tag{114}$$

From (112) we see that γ_{0m} is also a harmonic function, but it must satisfy the co-ordinate condition, i.e.

$$\underset{3}{\gamma_{0r|r}} = \underset{2}{\gamma}{}_{00|\underset{1}{0}} = \sum_{i=1}^{p} (4\underset{2}{\overset{i}{m}}\overset{i}{\psi}_{|r}\underset{1}{\overset{i}{\dot{\xi}}}_r) \tag{115}$$

since

$$\overset{i}{\psi}_{|0} = -\overset{i}{\psi}_{|k}\overset{i}{\dot{\xi}}{}^k, \qquad \overset{i}{\dot{\xi}}{}^k = \frac{d\overset{i}{\xi}{}^k}{d\tau} \tag{116}$$

The constant $\underset{2}{\overset{i}{m}}$ is identified as the (positive) gravitational mass. Thus

$$\underset{3}{\gamma_{0n}} = \sum_{i=1}^{p} 4\underset{2}{\overset{i}{m}}\overset{i}{\psi}\underset{1}{\overset{i}{\dot{\xi}}}{}^n \tag{117}$$

Consider now the case of two particles. We write (omitting the order index)

$$\phi = f+g, \qquad f = -2\overset{1}{m}\overset{1}{\psi}, \qquad g = -2\overset{2}{m}\overset{2}{\psi} \tag{118}$$

$$\overset{1}{\xi}_r = \eta_r, \qquad \overset{2}{\xi}_r = \zeta_r \tag{119}$$

The first of the surface integrals (111) vanishes for $l = 2$, because

$$\int_{\overset{i}{\Omega}} (2\underset{3}{\Lambda'_{0m}} - \underset{2}{\gamma}_{00|0|m}\underset{1}{)}n_m \, d\Omega = -\int_{\overset{i}{\Omega}} \underset{2}{\gamma}_{00|0|m}\underset{1}{n_m} \, d\Omega = 0 \tag{120}$$

So that, we are left with

$$\underset{4}{\overset{1}{C}_m} \equiv \frac{1}{4\pi}\int_{\overset{1}{\Omega}} 2\underset{4}{\Lambda_{mr}} n_r \, d\Omega = 0 \tag{121}$$

$$\underset{4}{\overset{2}{C}_m} \equiv \frac{1}{4\pi}\int_{\overset{2}{\Omega}} 2\underset{4}{\Lambda_{mr}} n_r \, d\Omega = 0 \tag{122}$$

These turn out to be

$$\begin{aligned} \underset{4}{\overset{1}{C}_m}(\tau) &\equiv 4\overset{1}{m}\{\ddot{\eta}^m + (1/2)\tilde{g}_{|m}\} = 0 \\ \underset{4}{\overset{2}{C}_m}(\tau) &\equiv 4\overset{2}{m}\{\ddot{\xi}^m + (1/2)\tilde{f}_{|m}\} = 0 \end{aligned} \tag{123}$$

$$\begin{aligned} \tilde{g}_{|m} &\equiv g_{|m} \quad \text{for} \quad x^s = \eta^s \\ \tilde{f}_{|m} &\equiv f_{|m} \quad \text{for} \quad x^s = \zeta^s \end{aligned} \tag{124}$$

The form of (123) is actually independent of the variables x^s. In fact

$$\tilde{g}_{|m} = \frac{\partial g(r)}{\partial \eta^m} = -\frac{\partial g(r)}{\partial \zeta^m}$$

$$g(r) = -\frac{2\overset{2}{m}}{r}, \qquad r^2 = (\eta^s - \zeta^s)(\eta^s - \zeta^s) \tag{125}$$

We can therefore think of our equations of motion as involving derivatives of functions of the position of the singularities. Equations (123) are the Newtonian equations of motion derived as the first approximation from the field equations.

This remarkable fact, that the equations of motion of particles can be derived from the free-field equations alone, is characteristic of the non-linear theory of gravitation. However, it should be emphasized that this procedure does not in anyway remove the field-particle duality inherent in the theory. *The introduction of singularities in the field to represent matter constitutes in effect a dualistic viewpoint.*

B. Quantum Theory

1. Quantum theory of a particle

Principle of indeterminacy. Classical physics is based on the metaphysical assumption that a physical system under consideration has an objective reality *independent* of the observer. In other words, the system is macroscopic in the sense that all disturbances in the system due to interaction of the system with the observer's measuring apparatus are negligible. Consequently, in classical mechanics it is meaningful to assign, for example, to a particle its position as well as momentum simultaneously at any given time. The whole objective of classical mechanics then reduces to the problem of specification of the forces involved and calculating the final position and momentum after a certain lapse of time from the equations of motion. It is well known that the final position and momentum of a particle are related to their initial values by means of *canonical transformations*.

Heisenberg showed that in a microscopic physical system it is no longer possible to ignore the disturbances in the system due to a measurement, as the disturbances themselves can be of the same order of magnitude as that of the quantity to be measured. By means of a series of Gedanken experiments he showed that the measurement of position of an electron, for example, entails an unpredictable change in its momentum; in particular if ΔQ and ΔP represent the uncertainties in the knowledge of position and momentum, respectively, one always has the uncertainty relations: $\Delta Q \, \Delta P \sim h$ (Planck's constant). Heisenberg's principle of indeterminacy asserts that at any given time it is possible to determine only one observable (and all other observables which are trivially related to it) with exactitude, the rest of the observables remaining completely indeterminate. The classical description of a particle moving along a path in the phase-space is therefore no longer tenable in microscopic physics. The new formalism which incorporates this fundamental principle of indeterminacy in a most satisfactory manner is called quantum mechanics. It consists, first of all, of an abstract mathematical structure, together with a correspondence between the mathematical objects and physical quantities and finally a set of physical axioms.

General formalism of quantum mechanics. (a) *Mathematical.* The relevant mathematical structure one must consider in quantum mechanics is that of Hilbert spaces.

A Hilbert space $\mathscr{H}$ is an abstract set of elements, called vectors, $|f\rangle$, $|g\rangle, \ldots$, etc., having the following set of properties:

I. $\mathscr{H}$ is an infinite dimensional linear vector space over the field of complex numbers. We denote the null-vector of $\mathscr{H}$ by $|0\rangle$.

II. There is defined a scalar product in $\mathscr{H}$, denoted by $(|f\rangle, |g\rangle)$ or $\langle f|g\rangle$ = a complex number, such that

(i) $(|f\rangle, \alpha|g\rangle) = \alpha(|f\rangle, |g\rangle)$, α = any complex number.
(ii) $(|f\rangle, |g_1\rangle + |g_2\rangle) = (|f\rangle, |g_1\rangle) + (|f\rangle, |g_2\rangle)$
(iii) $(|f\rangle, |g\rangle) = \overline{(|g\rangle, |f\rangle)}$; bar denotes complex conjugate.
(iv) $(|f\rangle, |f\rangle) \geqslant 0$ $(=0$ only when $|f\rangle = |0\rangle)$

from which follows

(v) $(\alpha|f\rangle, |g\rangle) = \bar{\alpha}(|f\rangle, |g\rangle) \equiv \bar{\alpha}\langle f|g\rangle$
(vi) $(|f_1\rangle + |f_2\rangle, |g\rangle) = \langle f_1|g\rangle + \langle f_2|g\rangle$
(vii) $\| |f\rangle\| \cdot \| |g\rangle\| \geqslant |\langle f|g\rangle|$ $(=\langle f|g\rangle$ only when $|f\rangle = \lambda|g\rangle)$

Here $\| |f\rangle\| = \sqrt{\langle f|f\rangle}$ = norm of $|f\rangle$. The above inequality is called Schwarz's inequality. We shall write $|f_n\rangle \to |f\rangle$ if $\| |f_n\rangle - |f\rangle\| \to 0$ as $n \to \infty$, i.e., the sequence of vectors $\{|f_n\rangle\}$ converge to $|f\rangle$. A sequence of vectors $\{|f_n\rangle\}$ is called a Cauchy sequence if, for any arbitrary real number $\varepsilon > 0$, there exists a number $N(\varepsilon)$ such that

$$\| |f_n\rangle - |f_m\rangle\| < \varepsilon \quad \text{whenever} \quad n, m > N(\varepsilon)$$

III. $\mathscr{H}$ is *complete*, i.e. for every Cauchy sequence $\{|f_n\rangle\}$ in $\mathscr{H}$ there exists a vector $|f\rangle$ in $\mathscr{H}$ such that $|f_n\rangle \to |f\rangle$.

Two vectors $|f\rangle, |g\rangle$ are called orthogonal if $\langle f|g\rangle = 0$. A system of vectors $\{|f_n\rangle\}$ is called an orthonormal system if $\langle f_n|f_m\rangle = \delta_{nm}$.

$\{|f_n\rangle\}$ is called a *complete orthonormal system* of $\mathscr{H}$ if for every $|f\rangle$ in $\mathscr{H}$ we have

$$|f\rangle = \sum_n^{\infty} \alpha_n |f_n\rangle, \quad \alpha_n \text{ complex numbers}$$

Then
$$\langle f_m|f\rangle = \sum_n^{\infty} \alpha_n \langle f_m|f_n\rangle = \alpha_m$$

so that
$$|f\rangle = \sum_n^{\infty} \langle f_n|f\rangle |f_n\rangle \tag{126}$$

$\{\langle f_n|f\rangle\}$ are called representatives of $|f\rangle$ relative to the complete orthonormal system $\{|f_n\rangle\}$.

A bounded linear operator A is defined as a mapping $A|f\rangle = |g\rangle$ of $\mathscr{H}$ into itself (or a subset of $\mathscr{H}$) such that

(i) $A(\alpha|f\rangle + \beta|g\rangle) = \alpha A|f\rangle + \beta A|g\rangle$
(ii) $\|A|f\rangle\| \leqslant C\| |f\rangle\|$ with constant C for all $|f\rangle$ in $\mathscr{H}$.

A bounded linear operator A is continuous in the sense if $|f_n\rangle \to |f\rangle$ then

$A\,|f_n\rangle \to A\,|f\rangle$. We say $A = B$ if $A\,|f\rangle = B\,|f\rangle$ for all $|f\rangle$ in $\mathscr{H}$. The definition of the following is quite obvious

(i) Identity operator I: $I\,|f\rangle = |f\rangle$ for all $|f\rangle$ in $\mathscr{H}$
(ii) Null operator 0: $0\,|f\rangle = |0\rangle$,, ,, ,, ,, ,,
(iii) Sum of A and B: $(A+B)\,|f\rangle = A\,|f\rangle + B\,|f\rangle$,, ,, ,, ,, ,,
(iv) Product of A and B: $AB\,|f\rangle = A(B\,|f\rangle)$,, ,, ,, ,, ,,

In general $AB \neq BA$; $[A, B] \equiv AB - BA$ is called the commutator of A and B.

Let A be a bounded linear operator in $\mathscr{H}$. Consider a fixed vector $|g\rangle$ in $\mathscr{H}$. Then $(|g\rangle, A\,|f\rangle)$, where $|f\rangle$ runs over all vectors in $\mathscr{H}$, is a bounded linear form in $|f\rangle$, written $L(|f\rangle) = (|g\rangle, A\,|f\rangle)$. For such a linear form there exists a unique $|g'\rangle$ in $\mathscr{H}$ such that $L(|f\rangle) = (|g\rangle, A\,|f\rangle) = (|g'\rangle, |f\rangle)$. The mapping $|g'\rangle = A^*\,|g\rangle$ is then defined for all $|g\rangle$ in $\mathscr{H}$ and A^*, called the *adjoint* of A, is also linear and bounded. Thus the adjoint A^* of a bounded linear operator A is a bounded linear operator such that $(|g\rangle, A\,|f\rangle) = (A^*\,|g\rangle, |f\rangle)$ for all $|g\rangle, |f\rangle$ in $\mathscr{H}$. The adjoint has the following properties:

(i) $(\alpha A)^* = \bar{\alpha}A^*$
(ii) $(A+B)^* = A^* + B^*$
(iii) $(AB)^* = B^*A^*$
(iv) $(A^*)^* = A$

A is called a Hermetian operator if $A = A^*$, i.e. $(|g\rangle, A\,|f\rangle) = (A\,|gf\rangle)$ for all $|g\rangle, |f\rangle$ in $\mathscr{H}$. Putting $|g\rangle = |f\rangle$ we get $(|f\rangle, A\,|f\rangle) = (A\,|f\rangle, |f\rangle) = (|f\rangle, A\,|f\rangle) =$ real.

If A is an operator and there exists a vector $|A'\rangle \neq |0\rangle$ such that $A\,|A'\rangle = A'\,|A'\rangle$, A' being complex number, we say that $|A'\rangle$ is an *eigenvector* of A corresponding to the *eigenvalue* A'. Hermetian operators have the following properties:

(i) The eigenvalues of a Hermetian operator are real.
(ii) If $|A'\rangle$ and $|A''\rangle$ be two eigenvectors of a Hermetian operator A corresponding to the eigenvalues $A' \neq A''$ then $|A'\rangle$ is orthogonal to $|A''\rangle$, i.e., $\langle A'|A''\rangle = 0$.
(iii) The eigenvectors of a bounded Hermetian operator after normalization form a denumerable complete orthonormal system. Consequently its eigenvalues form a discrete set (discrete spectrum).

The last property of bounded Hermetian operators is very important in quantum mechanics. Consider such an operator A and its normalized eigenvectors $|A'\rangle, |A''\rangle \ldots$, etc. An arbitrary vector $|\psi\rangle$ in $\mathscr{H}$ can then be expanded as

$$|\psi\rangle = \sum_{A'} \langle A'|\psi\rangle\,|A'\rangle \tag{127}$$

with

$$\langle A'|A''\rangle = \delta_{A'A''} \tag{128}$$

Every vector $|\psi\rangle$ of $\mathscr{H}$ can therefore be represented by its representative $\langle A'|\psi\rangle$, which is a function defined on the discrete spectrum of A. The scalar product of two vectors is then given by

$$\langle\psi|\phi\rangle = \sum_{A'} \langle\psi|A'\rangle\langle A'|\phi\rangle$$

In particular
$$\sum_{A'} \langle\psi|A'\rangle\langle A'|\psi\rangle = \| \, |\psi\rangle\|^2 < \infty$$

Conversely, every quadratically summable function $\langle A'|\psi\rangle$ represents a vector in a Hilbert space. The abstract Hilbert space $\mathscr{H}$ is therefore mapped on the Hilbert space $\mathscr{F}$ of quadratically summable functions on the spectrum of A. We call $\mathscr{F}$ the A-representation. The action of an operator B on $|\psi\rangle$, e.g. $B\,|\psi\rangle$ is then represented by

$$(|A'\rangle, B\,|\psi\rangle \equiv \langle A'|B|\psi\rangle = \sum_{A''} \langle A'|B|A''\rangle\langle A''|\psi\rangle$$

Thus the operator B is represented by the matrix $\langle A'|B|A''\rangle$. In $\mathscr{F}$ or A-representation the action of B can be written as

$$B\langle A'|\psi\rangle = \sum_{A''} \langle A'|B|A''\rangle\langle A''|\psi\rangle \tag{129}$$

A in its own representation is a diagonal matrix:

$$A\langle A'|\psi\rangle = A'\langle A'|\psi\rangle \tag{130}$$

The choice of a representation constitutes effectively a choice of a coordinate system. If one wishes to go from A-representation to B-representation, say, the relevant quantities in the new representation can be derived from the old ones by means of the so-called transformation functions $\langle A'|B'\rangle$. For example

$$\langle B'|\psi\rangle = \sum_{A'} \langle B'|A'\rangle\langle A'|\psi\rangle \tag{131}$$

$$\langle B'|C|B''\rangle = \sum_{A',\,A''} \langle B'|A'\rangle\langle A'|C|A''\rangle\langle A''|B''\rangle \tag{132}$$

In quantum mechanics one is sometimes forced to consider representations corresponding to operators which do not have a discrete spectrum. Unfortunately the above ideas do not carry over to the continuous case in a straightforward manner as there exist no proper eigenvectors corresponding to the continuous spectrum. One can, however, *formally* proceed in the continuous case with the help of improper eigenvectors by replacing the sum in (127) by an integral over the continuous spectrum σ.

$$|\psi\rangle = \int_{A'\varepsilon\sigma} \langle A'|\psi\rangle\,|A'\rangle\, dA' \tag{133}$$

The orthonormality condition (128) then must be replaced by

$$\langle A'|A''\rangle = \delta(A'-A'') \tag{134}$$

and the scalar product of two vectors becomes

$$\langle\psi|\phi\rangle = \int_{\sigma} \langle\psi|A'\rangle\langle A'|\phi\rangle\, dA'$$

The vectors $|\psi\rangle$ are then represented by quadratically integrable functions $\langle A'|\psi\rangle$ on the spectrum σ. The action of an operator B is then represented by a kernel function in the following way:

$$B\langle A'|\psi\rangle = \int_{\sigma} \langle A'|B|A''\rangle\langle A''|\psi\rangle\, dA'' \tag{135}$$

If A has a mixed spectrum (127) or (133) becomes

$$\begin{aligned}|\psi\rangle &= \sum_{A'} \langle A'|\psi\rangle\, |A'\rangle + \int_{A''\varepsilon\sigma} \langle A''|\psi\rangle\, |A''\rangle\, dA'' \\ &= \int_{(\sigma),\, A'} \langle A'|\psi\rangle\, |A'\rangle\end{aligned} \tag{136}$$

where the generalized notation includes summation over the discrete spectrum and integration over the continuous one. Self-adjoint operators having this property that any vector $|\psi\rangle$ in $\mathscr{H}$ can be expanded as in (127), (133) or (136) will be called "observable" operators.

Functions of observable operators can be defined in terms of the eigenvalue equation, i.e. if $A\,|A'\rangle = A'\,|A'\rangle, f(A)\,|A'\rangle = f(A')\,|A'\rangle$. In particular $A^{-1}\,|A'\rangle = A'^{-1}\,|A'\rangle$, with $A' \neq 0$. We have then $A^{-1}A = AA^{-1} = I$.

An operator U is called *unitary* if $U^{-1} = U^*$. Unitary operators have the following property. Consider a so-called unitary transformation:

$$\left.\begin{aligned}|\hat{A}'\rangle &= U\,|A'\rangle \\ \hat{A} &= UAU^*\end{aligned}\right\} \tag{137}$$

where A is an observable operator with eigenvectors $|A'\rangle$. Then $\|\,|\hat{A}'\rangle\| = \|\,|A\rangle\|$ and $\hat{A}U\,|A'\rangle = A'U\,|A'\rangle$, i.e. a unitary transformation leaves the norm of vectors and the eigenvalues invariant. Furthermore $\hat{A}\hat{B} = \widehat{(AB)}$ and $\hat{A}+\hat{B} = \widehat{(A+B)}$. Thus algebraic relations between operators remain unchanged under a unitary transformation.

An infinitesimal unitary transformation is of the form $U = I+i\varepsilon F$, with ε a real infinitesimal parameter and F self-adjoint, $F = F^*$. Under such a transformation an operator A becomes $\hat{A} = UAU^* \simeq A+i\varepsilon[F, A]$, so that the change is $\delta A = \hat{A}-A \simeq i\varepsilon[F, A]$.

A one-parameter group of unitary operators U_t (t a real parameter) is a family of unitary operators such that $U_{t_1}U_{t_2} = U_{t_1+t_2}$. Then $U_0 = I$ and $U_{-t} = (U_t)^{-1} = U_t^*$. Let $\lim_{\varepsilon\to 0}(U_\varepsilon - I)/\varepsilon$ exist and be equal to iA. Then $U_t = \exp iAt$, where A is self-adjoint, $A = A^*$.

General formalism of quantum mechanics. (b) *Physical.* The physical part of the general formalism of quantum mechanics consists first of all of the following *correspondence* between physical quantities and the mathematical objects defined above:

(i) The state of a physical system corresponds to a ray vector in a Hilbert space $\mathscr{H}$. This means $|\psi\rangle$ and $\lambda\,|\psi\rangle$, where λ is any non-zero complex number, represent the same state. In the following we shall assume the state vectors to be normalized to unity.

(ii) The dynamical observables of a physical system correspond to "observable" operators in $\mathscr{H}$.

We are now in a position to state the basic physical axioms.

Axiom I. The result of any measurement of an observable can only be one of the eigenvalues of the corresponding operator. As a result of the measurement, the physical system finds itself in the state represented by the corresponding eigenvector.

Axiom II. If a system is known to be in the state $|A'\rangle$, then the probability that a measurement of an observable B on the state $|A'\rangle$ yields the value B' is given by

$$w(A', B') = |\langle A'|B'\rangle|^2 \tag{138}$$

If B has a continuous spectrum, then $|\langle A'|B'\rangle|^2\,dB'$ gives the probability of B having the value in the range B' and $B'+dB'$.

Thus in every measurement the system makes a transition from an initial state $|A'\rangle$ to a final state $|B'\rangle$. Hence $w(A', B')$ is referred to as a transition probability.

Axiom III. The "observable" operators A, B satisfy the following commutation relations:

$$[A, B] = i\hbar\{A, B\}I \tag{139}$$

Here $\{A, B\}$ is the classical Poisson bracket of the corresponding dynamical observables and $2\pi\hbar$ is the Planck's constant. For example, the operators corresponding to the canonical position and momentum variables Q_i, P_i ($i = 1, 2, 3$) of a particle satisfy the fundamental commutation relations:

$$\left.\begin{aligned} [P_i, P_j] &= [Q_i, Q_j] = 0 \\ [Q_i, P_j] &= i\hbar\delta_{ij}I \end{aligned}\right\} \tag{140}$$

That the principle of indeterminacy is already embodied in the above formalism can be seen as follows. If $[A, B] \neq 0$, then the operators A, B do not have simultaneous eigenvectors; consequently they cannot be measured simultaneously with exactitude. We now give an exact formulation of the uncertainty relations.

We define the expectation value of an observable A on an arbitrary state $|\psi\rangle$ by

$$\langle A\rangle = \langle\psi|A|\psi\rangle \tag{141}$$

and the uncertainty ΔA of A by

$$\Delta A = \langle(A - I\langle A\rangle)^2\rangle^{1/2} \tag{142}$$

Then $(\Delta A)^2 = \langle A^2\rangle - \langle A\rangle^2$, so that if $|\psi\rangle = |A'\rangle$, $\Delta A = 0$. Moreover, $(\Delta A)^2$ is always $\geqslant 0$. Consider now two observable operators A, B and an arbitrary state $|\psi\rangle$. Write

$$\bar{A} = A - I\langle A\rangle$$

$$\bar{B} = B - I\langle B\rangle$$

so that $$(\Delta A)^2 = \langle\bar{A}^2\rangle, \qquad (\Delta B)^2 = \langle\bar{B}^2\rangle$$

and $$[\bar{A}, \bar{B}] = [A, B] = C, \quad \text{say.}$$

We have then

$$\langle\psi|C|\psi\rangle = \langle\psi|\bar{A}\bar{B}|\psi\rangle - \langle\psi|\bar{B}\bar{A}|\psi\rangle = (\bar{A}\,|\psi\rangle, \bar{B}\,|\psi\rangle) - (\bar{B}\,|\psi\rangle, \bar{A}\,|\psi\rangle)$$

If we put $(\bar{A}\,|\psi\rangle, \bar{B}\,|\psi\rangle) = x + iy$, where x, y are real we get

$$\langle C\rangle = i2y$$

Since $x^2 + y^2 \geqslant y^2$ we have

$$[\tfrac{1}{2}i\langle C\rangle]^2 \leqslant |(\bar{A}\,|\psi\rangle, \bar{B}\,|\psi\rangle)|^2$$

From Schwarz's inequality

$$\|\bar{A}\,|\psi\rangle\|^2\|\bar{B}\,|\psi\rangle\|^2 \geqslant |(\bar{A}\,|\psi\rangle, \bar{B}\,|\psi\rangle)|^2$$

Therefore finally $$\langle\bar{A}^2\rangle\langle\bar{B}^2\rangle \geqslant -\tfrac{1}{4}\langle C\rangle^2$$

or $$(\Delta A)^2(\Delta B)^2 \geqslant -\tfrac{1}{4}\langle[A, B]\rangle^2 \tag{143}$$

The Heisenberg uncertainty relation follows from (143) by putting

$$A = P, \quad B = Q; \qquad \Delta P\,\Delta Q \geqslant \hbar/2$$

So far we were concerned with states and observables at one instant of time. The dynamics of a system can be described in many ways. We discuss first of all the so-called *Schrödinger-picture*. In the Schrödinger-picture the state vectors are functions of time whereas the observable operators are independent of time. Let the state of a system at a time t_0 be $|\psi_{t_0}\rangle$, then the state at any other arbitrary time t, $|\psi_t\rangle$ can be written as

$$|\psi_t\rangle = U(t, t_0)|\psi_{t_0}\rangle \tag{144}$$

where $U(t, t_0)$ is an operator. The dynamics of the system are then described by the next axiom:

Axiom IV. $U(t, t_0)$ is a one-parameter group of unitary operators and has the form

$$U(t, t_0) = [\exp -(i/\hbar)H(t-t_0)] \tag{145}$$

Here H is the Hamiltonian operator of the system. $|\psi_t\rangle$ therefore satisfies the time-dependent Schrödinger's equation of motion

$$\frac{\partial|\psi_t\rangle}{\partial t} = -(i/\hbar)H\,|\psi_t\rangle \tag{146}$$

One can describe the dynamics of a system by an equivalent picture (*Heisenberg-picture*) by subjecting the vectors and operators in the Schrödinger-picture to the following unitary transformation:

$$\left.\begin{aligned} |\hat{\psi}_t\rangle &= U_t^{-1}\,|\psi_t\rangle \\ \hat{A}(t) &= U_t^{-1}AU_t \end{aligned}\right\} \tag{147}$$

where we have put $U_t = U(t, t_0)$. The effect of the above transformation is to make the state vectors stationary in time, as $|\hat{\psi}_t\rangle = |\psi_{t_0}\rangle$; and the operators become time-dependent. The equation of motion in the Heisenberg-picture becomes in view of (145) and (147):

$$i\hbar\dot{\hat{A}}(t) = [\hat{A}(t), \hat{H}(t)] \tag{148}$$

where the dot represents the time derivative and $\hat{H}(t) = U_t^{-1}HU_t$. If we now compare (148) with the classical equation of motion of a dynamical variable $A(t)$ in Poisson bracket form

$$\dot{A}(t) = \{A, H\} \tag{149}$$

and remember (139), we see that the Heisenberg's form of equation of motion is the exact quantum analogue of (149). We also see that unitary transformations play the same role in quantum mechanics as canonical transformations do in classical mechanics.

Quantum field-particle duality. We now illustrate the general formalism of quantum mechanics and show how the quantum field-particle duality arises by considering a one-dimensional harmonic oscillator whose classical Hamiltonian is given by

$$H = P^2/2m+(m\omega^2/2)Q^2 \tag{150}$$

Here m = mass of the oscillator, $\omega/2\pi$ = frequency of oscillation. The form of the Hamiltonian is taken over in quantum mechanics except that now P and Q are observable operators whose commutation relation is given by (140)

$$[P, Q] = (\hbar/i)I \tag{151}$$

We introduce the following operator

$$A = \frac{1}{\sqrt{2}}\left(\sqrt{\left(\frac{m\omega}{\hbar}\right)} Q + \frac{i}{\sqrt{(m\omega\hbar)}} P\right) \tag{152}$$

We have then from (150)-(152),

$$\left.\begin{aligned}&[A, A^*] = I\\ &H = \hbar\omega(N+\tfrac{1}{2}I) \quad \text{where} \quad N = A^*A\end{aligned}\right\} \tag{153}$$

We want to consider the N-representation which is the same as H-representation, since N commutes with H. The H-representation is also called the Heisenberg-representation. Let

$$N\,|N'\rangle = N'\,|N'\rangle; \qquad |N'\rangle \neq |0\rangle$$

or

$$A^*A\,|N'\rangle = N'\,|N'\rangle \tag{154}$$

Multiplying by A and using the commutator $[A, A^*]$, we have

$$(N+I)\,A\,|N'\rangle = N'\,A\,|N'\rangle$$

That is, either $A\,|N'\rangle$ is an eigenvector of N corresponding to the eigenvalue $N'-1$, or $A\,|N'\rangle = |0\rangle$. Similarly by multiplying (154) with A^* we deduce that, either $A^*\,|N'\rangle$ is an eigenvector of N corresponding to the eigenvalue $N'+1$ or $A^*\,|N'\rangle = |0\rangle$.

But $A^*\,|N'\rangle \neq |0\rangle$, because $\|A^*\,|N'\rangle\|^2 = (A^*\,|N'\rangle, A^*\,|N'\rangle) = (|N'\rangle, AA^*\,|N'\rangle) = (|N'\rangle, A^*A\,|N'\rangle) + \|\,|N'\rangle\|^2 = \|A\,|N'\rangle\|^2 + \|\,|N'\rangle\|^2 \neq 0$, since $|N'\rangle \neq |0\rangle$. Thus $A^*\,|N'\rangle$ is always an eigenvector of N corresponding to the eigenvalue $N'+1$.

The eigenvalues of N cannot be negative, because $N'\|\,|N'\rangle\|^2 = (|N'\rangle, N\,|N'\rangle) = (|N'\rangle, A^*A\,|N'\rangle) = \|A\,|N'\rangle\|^2$ or $N' = \|A\,|N'\rangle\|^2/\|\,|N'\rangle\|^2 \geqslant 0$. We thus infer that there exists an eigenvector $|0'\rangle$ such that $A\,|0'\rangle = |0\rangle$; otherwise a repeated application of A on $|N'\rangle$ would lead us to a negative eigenvalue N'. Since $N\,|0'\rangle = 0\,|0'\rangle$, $|0'\rangle$ is an eigenvector of N corresponding to the eigenvalue zero. We can also conclude that the possible eigenvalues of N are 0, 1, 2, 3, etc. Consequently the eigenvalues of H are given by

$$H' = \hbar\omega(N'+\tfrac{1}{2}), \qquad N' = 0, 1, 2, 3, \ldots, \text{etc.} \tag{155}$$

The various eigenvectors of N can be built up from $|0'\rangle$ by repeated application of A^*, i.e. $|N'\rangle = (A^*)^{N'}\,|0'\rangle$. The complete orthonormal system of vectors which characterizes the H- or N-representation is given by $\{|\phi_{N'}\rangle\}$ where $|\phi_{N'}\rangle = |N'\rangle/\sqrt{N'!}$. According to (127) an arbitrary vector $|\psi\rangle$ is given by

$$|\psi\rangle = \sum_{N'} \langle\phi_{N'}|\psi\rangle|\phi_{N'}\rangle \tag{156}$$

One can now write down the matrix-representation of the operators P, Q, H.

$$\langle\phi_{N'}|H|\phi_{N''}\rangle = \hbar\omega(N''+\tfrac{1}{2})\,\delta_{N'N''} \tag{157}$$

$$\langle\phi_{N'}|P|\phi_{N''}\rangle = \frac{1}{i}\sqrt{\left(\frac{m\omega\hbar}{2}\right)}\left(\sqrt{(N'')}\,\delta_{N',N''-1}-\sqrt{(N''+1)}\,\delta_{N',N''+1}\right) \tag{158}$$

$$\langle\phi_{N'}|Q|\phi_{N''}\rangle = \sqrt{\left(\frac{\hbar}{2m\omega}\right)}\left(\sqrt{(N'')}\,\delta_{N',N''-1}+\sqrt{(N''+1)}\,\delta_{N',N''+1}\right) \tag{159}$$

According to (129) the action of the operator Q in the Heisenberg-representation is then given by

$$Q\langle\phi_{N'}|\psi\rangle = \sqrt{\left(\frac{\hbar}{2m\omega}\right)}\left(\sqrt{(N'+1)}\langle\phi_{N'+1}|\psi\rangle+\sqrt{(N')}\langle\phi_{N'-1}|\psi\rangle\right) \tag{160}$$

We now want to go over to the Q-representation. Our problem is therefore to find the transformation functions $\langle\phi_{N'}|Q'\rangle$. The Q-representation is also called the Schrödinger-representation. Q, being the position operator, must have a continuous spectrum, i.e.

$$Q\,|Q'\rangle = Q'\,|Q'\rangle \tag{161}$$

where

$$\langle Q'|Q''\rangle = \delta(Q'-Q'') \tag{162}$$

Consider now the eigenvalue equation (161) in the Heisenberg-representation, i.e.,

$$Q\langle\phi_{N'}|Q'\rangle = Q'\langle\phi_{N'}|Q'\rangle \tag{163}$$

We already know from (160) how Q operates on the representative of an arbitrary vector in the Heisenberg-representation. So, putting $|\psi\rangle = |Q'\rangle$ in (160) we have from (160) and (163)

$$Q'\langle\phi_{N'}|Q'\rangle = \sqrt{\left(\frac{\hbar}{2m\omega}\right)}\left(\sqrt{(N'+1)}\,\langle\phi_{N'+1}|Q'\rangle+\sqrt{(N')}\,\langle\phi_{N'-1}|Q'\rangle\right)$$

This recursion formula for the transformation functions is easily seen to be satisfied by

$$\langle\phi_{N'}|Q'\rangle = \frac{f\left[\sqrt{\left(\frac{m\omega}{\hbar}\right)}\,Q'\right]H_{N'}\left[\sqrt{\left(\frac{m\omega}{\hbar}\right)}\,Q'\right]}{\sqrt{(2^{N'}N'!)}} \tag{164}$$

where $H_{N'}$ are Hermite polynomials and the function f is determined by the orthonormality condition.

$$\sum_{N'} \langle Q'|\phi_{N'}\rangle\langle\phi_{N'}|Q''\rangle = \langle Q'|Q''\rangle = \delta(Q'-Q'')$$

f turns out to be:
$$f\left[\sqrt{\left(\frac{m\omega}{\hbar}\right)}Q'\right] = \frac{e^{-(\alpha Q')^2/2}}{\sqrt{\beta}}$$

where we have put

$$\alpha = \sqrt{\frac{m\omega}{\hbar}}, \qquad \beta = (\pi\alpha^{-2})^{1/2}$$

With the help of the transformation functions we can now express everything in the Schrödinger-representation. For example, according to (131) an arbitrary vector $|\psi\rangle$ has the representative $\langle Q'|\psi\rangle$ where

$$\langle Q'|\psi\rangle = \sum_{N'} \langle Q'|\phi_{N'}\rangle\langle\phi_{N'}|\psi\rangle$$

The operator Q in its own representation has obviously the following matrix representation

$$\langle Q'|Q|Q''\rangle = Q''\,\delta(Q'-Q'')$$

so that
$$Q\langle Q'|\psi\rangle = Q'\langle Q'|\psi\rangle \tag{165}$$

To calculate $\langle Q'|P|Q''\rangle$ we use the transformation formula (132):

$$\langle Q'|P|Q''\rangle = \sum_{N',N''} \langle Q'|\phi_{N'}\rangle\langle\phi_{N'}|P|\phi_{N''}\rangle\langle\phi_{N''}|Q''\rangle \tag{166}$$

Using (158) and the property of the Hermite polynomials it can be shown that

$$\sum_{N'} \langle Q'|\phi_{N'}\rangle\langle\phi_{N'}|P|\phi_{N''}\rangle = \frac{\hbar}{i}\frac{d}{dQ'}\langle Q'|\phi_{N''}\rangle \tag{167}$$

so that
$$\langle Q'|P|Q''\rangle = \sum_{N''}\frac{\hbar}{i}\frac{d}{dQ'}\langle Q'|\phi_{N''}\rangle\langle\phi_{N''}|Q''\rangle$$
$$= \frac{\hbar}{i}\frac{d}{dQ'}\delta(Q'-Q'')$$

Consequently
$$P\langle Q'|\psi\rangle = \frac{\hbar}{i}\frac{d}{dQ'}\langle Q'|\psi\rangle \tag{168}$$

(165) and (168) show how the position and momentum operators operate in the Schrödinger-representation. These results, although derived by considering the particular example of a harmonic oscillator, are quite obviously

completely general. The Hamiltonian operator, for example, for any one-dimensional system in the Schrödinger-representation is given by

$$H(P, Q)\langle Q'|\psi\rangle = H\left(\frac{\hbar}{i}\frac{d}{dQ'}, Q'\right)\langle Q'|\psi\rangle \tag{169}$$

Consider now the eigenvalue equation of H, i.e.,

$$H(P, Q)\,|H'\rangle = H'\,|H'\rangle$$

In the Schrödinger-representation it then becomes

$$H\left(\frac{\hbar}{i}\frac{d}{dQ'}, Q'\right)\langle Q'|H'\rangle = H'\langle Q'|H'\rangle \tag{170}$$

This equation is the so-called Schrödinger's wave equation. If one is interested in the dynamics of the system one must consider $|H', t\rangle = \exp\left[-(i/\hbar)Ht\right]$ $|H'\rangle$ (putting $t_0 = 0$; $|H'\rangle = |H', 0\rangle$). For a free particle, for example, the functions $\psi_{H'}(Q', t) \equiv \langle Q'|H', t\rangle$ describe plane waves in space-time. The states of a dynamical system are therefore represented by wavelike functions in the Schrödinger-representation. This is the origin of quantum field-particle duality. The physical significance of the wave functions must, however, be clearly kept in mind. $|\langle Q'|H', t\rangle|^2\, dQ'$ represents the probability of finding the particle in the region Q' to $Q'+dQ'$ if its energy is known to be H' at a time t. $\Psi_{H'}(Q', t)$ is therefore a probability field and not a real physical field in the classical sense.

One cannot fail to notice a certain methodological similarity between the theory of gravitation and quantum mechanics. The leitmotif behind both the theories is a universal physical principle—the principle of equivalence and the principle of indeterminacy. However, these physical principles do not by themselves imply the necessity of employment of the basic mathematical formalisms, e.g., Riemannian geometry and Hilbert spaces. It is just that these mathematical formalisms incorporate the respective physical principles in a most satisfactory manner. Moreover, both these formalisms imply a certain covariance—general covariance of co-ordinates in the theory of gravitation and complete equivalence of representations in quantum mechanics. Apart from aesthetics, the main justification for the employment of the respective mathematical formalisms lies in their verifiable predictions.

Part 2

Non-dualistic Theories

A. Classical Theories

B. Quantum Theory

A. Classical Theories

1. Field formalism

(a) *Electrodynamics of Mie.* We have seen that attempts to derive the inertial properties of charged particles from their self-field alone led to an impasse in the Maxwell-Lorentz theory. G. Mie was the first to propose a tentative non-dualistic field theory in which the charge and mass of particles could be determined solely from the field concept.

Consider the variational principle

$$\delta S = 0; \qquad S = \int L \, dx \tag{1}$$

We are once again in the Minkowski space-time.† Let the invariant Lagrangian L depend either explicitly on the potential 4-vector A_μ or through the field tensor $f_{\mu\lambda} = A_{\lambda,\mu} - A_{\mu,\lambda}$:

$$L \equiv L(f_{\mu\lambda}; A_\nu) \tag{2}$$

Then
$$\delta S = \int \delta L \, dx = \int (\tfrac{1}{2} d_{\mu\lambda} \, \delta f_{\mu\lambda} - j_\nu \, \delta A_\nu) \, dx$$

where we have put

$$d_{\mu\lambda} \equiv \frac{\partial L}{\partial f_{\mu\lambda}}, \qquad j_\nu \equiv -\frac{\partial L}{\partial A_\nu} \tag{3}$$

Therefore

$$\delta S = -\int (d_{\mu\lambda} \, \delta A_\lambda)_{,\mu} \, dx + \int (d_{\mu\lambda,\lambda} - j_\mu) \, \delta A_\mu \, dx$$

The first term, being a divergence, can be converted into a surface integral and therefore vanishes because $\delta A_\mu = 0$ at the boundary. So that $\delta S = 0$ implies the Euler-Lagrange equation:

$$\left.\begin{aligned} d_{\mu\lambda,\lambda} &= j_\mu \\ \text{or} \qquad \left(\frac{\partial L}{\partial f_{\mu\lambda}}\right)_{,\lambda} &= -\frac{\partial L}{\partial A_\mu} \end{aligned}\right\} \tag{4}$$

† Here $x^\mu = (\mathbf{x}, x^4 = ict)$.

The invariance of (1) under co-ordinate transformations furnishes us an energy-momentum tensor $T_{\mu\nu}$:

$$T_{\mu\nu,\nu} = 0, \qquad T_{\mu\nu} = 2d_{\mu\lambda}f_{\nu\lambda}+j_\mu A_\nu-\delta_{\mu\nu}L \tag{5}$$

If one takes $L = \lambda f_{\mu\lambda}f_{\mu\lambda}$ where λ is a constant factor of proportionality, we get $d_{\mu\lambda} \propto f_{\mu\lambda}, j_\mu = 0$, so that (4) reduces to the Maxwell's equation for the free-field case.

Mie takes $L = \lambda f_{\mu\lambda}f_{\mu\lambda}-f[\sqrt{(-A_\mu A_\mu)}]$ so that $d_{\mu\lambda}$ is still proportional to $f_{\mu\lambda}$ but $j_\mu \neq 0$, *which can be interpreted as the current* 4-*vector*. Mie's choice of L is of course entirely arbitrary because it is possible to form a number of other invariants from $f_{\mu\lambda}$ and A_μ. It is impossible to narrow down the number of alternatives to such an extent so as to lead unambiguously to a unique action function.

Mie assumes the field of a stationary charged particle to be static and spherically symmetric. We have then $\mathbf{A} = 0$, $A_4 = i\phi(r)$; $\mathbf{j} = 0$, $j_4 = ic\rho$ (here ρ is the charge density). From (3) we have

$$ic\rho = j_4 = -\frac{\partial L}{\partial A_4} \quad \text{or} \quad \rho = -\frac{f'(\phi)}{c} \tag{6}$$

The field-equations (4) become (apart from a constant factor)

$$\frac{1}{r^2}\frac{d}{dr}(r\phi'(r)) = -\frac{f'(\phi)}{c} \tag{7}$$

The function f is assumed to be such that (7) has *regular solutions everywhere including* $r = 0$. From (6) the total charge of the particle is given by

$$\theta = \int \rho r^2 \sin\theta \, dr \, d\theta \, d\phi = -\frac{4\pi}{c}\int r^2 f'(\phi)\, dr = -\frac{4\pi}{c}\left[r^2\phi'(r)\right]_0^\infty \tag{8}$$

The mass of the particle on the other hand is given by the energy-momentum tensor,

$$m = \frac{1}{c^2}\int T_{44}\, dr = \frac{1}{c^2}\int r^2\left[\frac{\phi}{2}f'(\phi)-f(\phi)\right] dr \tag{9}$$

(apart from a constant factor).

The mass and charge of a particle are therefore determined in the static case completely by the potential ϕ. According to Mie the particle is not localized at a point but is a characteristic of the field. Thus the field-particle duality is completely removed. The equation of motion of the particle is contained in (5).

One can make two principal objections to Mie's theory. A Lagrangian which is to correspond to reality will have to lead to one and one only solution for every kind of particle. It has not been possible to find a Lagrangian which satisfies this condition.

Second if ϕ is a solution for the electrostatic potential of a particle, $\phi+\text{const}$ is no longer a solution as the field-equation (7) depends explicitly on the absolute value of the potential. A material particle will therefore not be able to exist in a constant external potential. This is also exhibited by the fact that Mie's Lagrangian is not invariant under gauge-transformations.

In spite of these objections it was Mie who first showed the usefulness of constructing a generalized electrodynamics based only on the various invariants of the electromagnetic field. Also for the first time an attempt was made to derive the characteristics of particles from that of fields.

(b) *Theory of Born and Infeld.* Born and Infeld's Lagrangian is a function of $f_{\mu\lambda}$ only, but it is no longer a linear function of $f_{\mu\lambda}f_{\mu\lambda}$. Suppose we take

$$L \equiv L(f_{\mu\lambda}f_{\mu\lambda})$$

Then we have the field equations:

$$\left.\begin{aligned} d_{\mu\lambda,\lambda} &= 0 \\ d_{\mu\lambda} &\equiv \frac{\partial L}{\partial f_{\mu\lambda}} \end{aligned}\right\} \tag{10}$$

For an arbitrary function $L(f_{\mu\lambda}f_{\mu\lambda})$, $d_{\mu\lambda}$ is no longer in general proportional to $f_{\mu\lambda}$ as in the Maxwell's theory. One can then *define* $j_\mu \equiv (c/4\pi)f_{\mu\lambda,\lambda}$ as the current-vector.

Born and Infeld consider the following invariants which can be formed from $f_{\mu\lambda}$:

$$\left.\begin{aligned} L_f &= -\frac{1}{16\pi}f_{\mu\lambda}f_{\mu\lambda} = \frac{1}{8\pi}(E^2-H^2) \\ L_f' &= \tfrac{1}{2}\varepsilon^{\alpha\beta\gamma\delta}f_{\alpha\beta}f_{\gamma\delta} = (\mathbf{E}.\mathbf{H}) \end{aligned}\right\} \tag{11}$$

Here L_f is the free-field Lagrangian of Maxwell, $\varepsilon^{\alpha\beta\gamma\delta}$, the completely anti-symmetric Levi-Civita symbol. The Born-Infeld Lagrangian is

$$L = \frac{E_0^2}{4\pi}\left[1-\sqrt{\left(1-\frac{E^2-H^2}{E_0^2}-\frac{(\mathbf{E}.\mathbf{H})}{E_0^4}\right)}\right] \tag{12}$$

Here E_0 is a constant whose physical significance is that of a maximal field. For E/E_0, $H/E_0 \ll 1$, L reduces to the free-field Lagrangian L_f.

Consider now a static spherically symmetric field. We set $\mathbf{H} = 0$, $\mathbf{E} = -\text{grad}\,\phi$. Then

$$L = \frac{E_0^2}{4\pi}\left[1-\sqrt{\left(1-\frac{E^2}{E_0^2}\right)}\right] \tag{13}$$

The equations corresponding to (10) are then

$$-\frac{1}{4\pi}\frac{\partial D_k}{\partial x_k} = 0 \tag{14}$$

where
$$D_k = 4\pi\frac{\partial L}{\partial E_k} = \frac{E_k}{\sqrt{(1-E^2/E_0^2)}} \tag{15}$$

D corresponds to the field $d_{\mu\lambda}$.

The solution of (14) is $\mathbf{D} = e\mathbf{r}/r^3$, e = const of integration. From (15)

$$\mathbf{E} = \frac{\mathbf{D}}{\sqrt{(1+D^2/E_0^2)}} = \frac{e\mathbf{r}}{r\sqrt{(r^4+r_0^4)}} \tag{16}$$

where $r_0 = \sqrt{(e/E_0)}$. The potential ϕ is then given by

$$\phi = \int_r^\infty E_r\,dr = \frac{e}{r_0}\int_{r/r_0}^\infty \frac{dx}{\sqrt{(1+x^4)}}$$

At the origin the electric field becomes

$$\left.\begin{aligned} E(0) &= \frac{e}{r_0^2} = E_0 \\ \text{and}\qquad \phi(0) &= \frac{e}{r_0}\int_0^\infty \frac{dx}{\sqrt{(1+x^4)}} = (1{\cdot}8541\ldots)\frac{e}{r_0} \end{aligned}\right\} \tag{17}$$

The charge-density according to the definition is

$$\rho = \frac{\operatorname{div}\mathbf{E}}{4\pi} = \frac{er_0^4}{2\pi r(r^4+r_0^4)^{3/2}} \tag{18}$$

and the total charge is given by

$$\int \rho\,d\mathbf{r} = \int \frac{er_0^4\,d\mathbf{r}}{2\pi r(r^4+r_0^4)^{3/2}} = e$$

From the standpoint of the **D**-field the charge thus behaves as a point particle. On the other hand, from the standpoint of the electric field **E** the charge is distributed in an *effective radius* r_0; ρ goes to zero for $r \gg r_0$.

We consider now the energy-momentum tensor. From (5) and (13) we have

$$T_{44} = \frac{1}{4\pi}(\mathbf{E}.\mathbf{D}) - L \tag{19}$$

so that the mass of the particle is given by

$$m = \frac{1}{c^2}\int T_{44}\,d\mathbf{r} = \frac{2}{3}\frac{e\phi_0}{c^2} = (1{\cdot}2361\ldots)\frac{e^2}{c^2 r_0} \tag{20}$$

We thus have a non-dualistic and non-linear field theory which is formally free from objections. The field energy is finite and the energy-momentum of the field has the correct transformation properties. So Born-Infeld succeeds where the classical electromagnetic mass theory of Abraham failed. In spite of its formal success the theory cannot explain, for example, the spin of the electron. Moreover, it does not lead to anything new even in its quantized version.

2. *Particle formalism*

(a) *Electrodynamics of Wheeler and Feynman.* It can be argued that the classical field is a *secondary* concept and that it should be possible to reconstruct classical electrodynamics on the primary concept of charged particles only. Wheeler and Feynman have shown that one can revive the action-at-a-distance principle in electrodynamics and reproduce all the basic aspects of the field-particle theory of Maxwell-Lorentz in a *pure particle* formalism.

Consider a system of particles with mass $m_{(a)}$, charge $e_{(a)}$ and position co-ordinates $\xi_\mu^{(a)}$. We shall now use real co-ordinates $\xi_\mu \equiv (\boldsymbol{\xi}, c\tau)$ for the background Minkowski space-time so that the metric tensor is $g_{11} = g_{22} = g_{33} = -g_{44} = 1$. Let the co-ordinates $\xi_\mu^{(a)}$ of the ath particle depend on the ath "proper co-time" α ($\alpha = c\tau$ in a rest frame); $\xi_\mu^{(b)}$ depend on the bth "proper co-time" β and so on. We denote $d\xi_\mu^{(a)}/d\alpha$ by $\dot{\xi}_\mu^{(a)}$.

The motion of each particle in such a system is assumed to follow from the variational principle:

$$\delta S = 0, \qquad S = -\sum_a m_{(a)}c \int (-\dot{\xi}_\mu^{(a)}\dot{\xi}^{(a)\mu})^{1/2}\, dx$$
$$+ \sum_{a<b} \frac{e_{(a)}e_{(b)}}{c} \iint \delta[(\xi_\mu^{(a)} - \xi_\mu^{(b)})(\xi^{(a)\mu} - \xi^{(b)\mu})]\dot{\xi}_\mu^{(a)}\dot{\xi}^{(b)\mu}\, d\alpha\, d\beta \tag{21}$$

Here we have a δ-function under the second integral which represents the interaction between the pairs of particles. The equation of motion of the ath particle is to follow as before from the variation of $\xi_\mu^{(a)}$. Let us *denote*

$$A_\mu^{(a)}(x) = e_{(a)} \int \delta[(x_\mu - \xi_\mu^{(a)})(x^\mu - \xi^{(a)\mu})]\dot{\xi}_\mu^{(a)}\, d\alpha \tag{22}$$

Then we have

$$A_{\mu,\mu}^{(a)} = 0 \tag{23}$$

Equation (21) implies, for a variation of $\xi_\mu^{(a)}$, the following equation of motion for the ath particle:

$$m_{(a)}c^2\ddot{\xi}_\mu^{(a)} = e_{(a)}\left[\sum_{b\neq a} f_{\mu\lambda}^{(b)}(\xi^a)\right]\dot{\xi}_\lambda^{(a)} \tag{24}$$

where we have put

$$f_{\mu\lambda}^{(b)}(x) = A_{\lambda,\,\mu}^{(b)}(x) - A_{\mu,\,\lambda}^{(b)}(x) \tag{25}$$

We thus obtain the Lorentz equation of motion in terms of $\sum_{b\neq a} f_{\mu\lambda}^{(b)}$, defined by (23) and (25), which can therefore be interpreted as the "field" acting on the ath particle. The right-hand side of (24), however, *does not contain* any self-field term.

It can now be shown that $f_{\mu\lambda}^{(a)}(x)$ obeys Maxwell's equation. We use the relation:

$$\delta(\partial^2/\partial x_\mu\,\partial x^\mu)\,\delta[(x_\mu - \xi_\mu^{(a)})(x^\mu - \xi^{(a)\mu})] = -4\pi\delta(x - \xi^{(a)}) \tag{26}$$

Then

$$(\partial^2/\partial x_\mu\,\partial x^\mu)A_\mu^{(a)}(x) = -4\pi j_\mu^{(a)}(x) \tag{27}$$

where

$$j_\mu^{(a)}(x) = e_{(a)}\int \delta(x - \xi^{(a)})\dot{\xi}_\mu^{(a)}\,d\alpha \tag{28}$$

From (25) and (27) we get

$$f_{\mu\lambda,\,\lambda}^{(a)}(x) = 4\pi j_\mu^{(a)}(x) \tag{29}$$

The physical significance of $A_\mu^{(a)}(x)$ is that it is the sum of the usual half-retarded and half-advanced potentials of the Maxwell-Lorentz theory due to the ath particle.

$$A_\mu^{(a)}(x) = \tfrac{1}{2}A_{\mu\,\mathrm{ret}}^{(a)}(x) + \tfrac{1}{2}A_{\mu\,\mathrm{adv}}^{(a)}(x) \tag{30}$$

We thus have a pure particle theory where the notion of field does not enter as an independent entity; the motion of a given particle is completely determined by the sum of "fields" $f_{\mu\lambda}^{(b)}(x)$ produced by every particle *other* than the given particle. There is no action of a charge upon itself; consequently there is no self-field or self-energy problem.

(b) *Whitehead's theory of gravitation.*† Shortly after Einstein, Whitehead formulated a pure particle theory of gravitation which bears a striking resemblance to the electrodynamics of Wheeler and Feynman.

† We follow here the lucid presentation of J. L. Synge: The relativity theory of A. N. Whitehead. Lecture Series No. 5, Inst. for Fluid Dynamics and Applied Mathematics, University of Maryland, 1951.

Whitehead's theory is also set in the background of the Minkowski space-time. We employ now the imaginary co-ordinates $x_\mu = x^\mu = (\mathbf{x}, ict)$. Let there be a particle of proper mass m moving along the world-line L' given by $x'_\mu = x'_\mu(s')$, where s' is the arc length, i.e., $ds'^2 = -dx'_\mu\, dx'_\mu$. Let the unit tangent vector to L' be denoted by $\lambda'_\mu = dx'_\mu/ds'$.

According to Whitehead the gravitational "field" at a point x_μ due to the particle is given by a symmetric tensor $g_{\mu\lambda}$ defined as follows:

$$\left.\begin{aligned} g_{\mu\lambda} &= \delta_{\mu\lambda}+\tilde{g}_{\mu\lambda} \\ \text{where} \qquad \tilde{g}_{\mu\lambda} &= (mk/\omega^3)\xi_\mu\xi_\lambda \end{aligned}\right\} \tag{31}$$

and
$$\xi_\mu = x_\mu - x'_\mu \tag{32}$$

Here x'_μ is the event where L' is cut by the null-cone

$$(x_\mu - x'_\mu)(x_\mu - x'_\mu) = 0 \tag{33}$$

drawn into the past from x_μ, and

$$\omega = -\xi_\mu\lambda'_\mu \tag{34}$$

and $$k = 2G/c^3, \qquad G = \text{constant of gravitation}$$

If we have a system of particles with masses $m_{(a)}$ $(a = 1, 2, \ldots)$ and world-lines $L^{(a)}$, the total "field" is then defined to be

$$\left.\begin{aligned} g_{\mu\lambda} &= \delta_{\mu\lambda}+\sum_a \tilde{g}^{(a)}_{\mu\lambda} \\ \text{where} \qquad \tilde{g}^{(a)}_{\mu\lambda} &= (m_{(a)}k/\omega^{(a)^3})\xi^{(a)}_\mu\xi^{(a)}_\lambda \\ \omega^{(a)} &= -\xi^{(a)}_\mu\lambda^{(a)}_\mu \end{aligned}\right\} \tag{35}$$

(no summation over a).

Thus, as in Einstein's theory, the gravitational "field" is described by a symmetric tensor $g_{\mu\lambda}$, which has, however, nothing to do with the geometry of space-time and as in Wheeler-Feynman's theory is only an auxiliary concept.

Consider a particle at rest with the time-axis coincident with its world-line. We have then

$$\left.\begin{aligned} &\lambda'_k = 0 \quad (k = 1, 2, 3), \qquad \lambda'_4 = i \\ &\xi_k = x_k; \qquad \xi_4 = ir, \quad \omega = r \end{aligned}\right\} \tag{36}$$

where r is the spatial distance of the point of observation P from the particle for which $x'_k = 0$. The "field" due to such a particle at rest is then from above

$$\left.\begin{aligned} g_{kl} &= \delta_{kl}+\left(\frac{mk}{r^3}\right)x_k x_l \\ g_{k4} &= i(mk/r^2)x_k \\ g_{44} &= 1-(mk/r) \end{aligned}\right\} \tag{37}$$

Consider now the fundamental form

$$\begin{aligned} \phi &= g_{\mu\lambda}\, dx_\mu\, dx_\lambda \\ &= dx_k\, dx_k+(mk/r^3)(x_k\, dx_k)^2+2i(mk/r^2)x_k\, dx_k\, dx_4+(1-mk/r)\, dx_4^2 \end{aligned}$$

In polar co-ordinates the Whitehead fundamental form for the "field" of a particle at rest becomes

$$\begin{aligned} \phi = (1+mk/r)\, dr^2+r^2(d\theta^2+\sin^2\theta\, d\phi^2)-2(mk/r)\, dr\, c\, dt \\ -(1-mk/r)c^2\, dt^2 \end{aligned} \tag{38}$$

Eddington† showed that the Whitehead form (38) can be transformed into the Schwarzschild form 1 (77) in general relativity by the following transformation of (r, θ, ϕ, t) to (r, θ, ϕ, τ):

$$ct = c\tau+f(r) \tag{39}$$

where $$f'(r) = -\frac{mk}{r-mk}$$

According to Whitehead the motion of a particle in the presence of "field" due to other particles is given by the variational principle

$$\delta\int d\bar{s} = 0 \tag{40}$$

where $$d\bar{s} = -\bar{g}_{\mu\lambda}\, dx_\mu\, dx_\lambda$$

$\bar{g}_{\mu\lambda}$ being the same as $g_{\mu\lambda}$ of (35) except that *the "self-field" due to the particle itself is omitted in the sum.*

In view of the transformation (39) and the equation of motion (40), Whitehead's theory makes almost the same formal predictions as in Einstein's theory of gravitation.

(c) *Theory of Hoyle and Narlikar.* Whereas Whitehead's theory is set in the background of Minkowski space-time, Hoyle and Narlikar accept the Riemannian space-time as the appropriate geometry. Their approach to the problem of gravitation is, however, quite different from that of Einstein.

† Eddington, A. S. (1914) *Nature* **113**, 192.

Consider the action function in Einstein's theory of gravitation for a system of particles

$$S = \frac{1}{16\pi G}\int R\sqrt{(-g)}\,dx - \sum_a m_a \int da \tag{41}$$

with

$$\delta S = 0$$

from which follow the field-equations

$$R_{\mu\lambda} - \tfrac{1}{2}g_{\mu\lambda}R = -8\pi G T_{\mu\lambda} \tag{42}$$

The first term in (41) is responsible for the left-hand side of (42) whereas the second term contributes the energy-momentum tensor $T_{\mu\lambda}$. It is a characteristic feature of all non-dualistic theories that the action function always contains two dissimilar integrals—a volume integral for the field variables, and a line integral for the particle variables. We have seen in Wheeler-Feynman's theory that in electromagnetism it is possible to convert the volume integral involving field variables into a double line integral over the world-lines of the particles. Hoyle and Narlikar go one step further, that is, to fuse the two terms in (41) into one double line-integral. This is done by invoking the Mach's principle, namely, *the mass m_a of the particle "a" should arise from all other particles of the universe*. One defines a mass function at a general point x due to particle a by

$$m^{(a)}(x) = -\lambda \int G(x, A)\,da \tag{43}$$

where λ is a coupling constant and $G(x, A)$, a symmetric scalar Green's function to be specified later. The world-lines of particles and their respective proper-times are denoted by the same label $a, b, c, \ldots$. A typical point on a will be denoted by A and a general space-time point will be denoted by x. Then the mass m_a of the particle a is given by

$$m_a = \sum_{b\neq a} m^{(b)}(A) = -\lambda \sum_{b\neq a} \int G(A, B)\,db \tag{44}$$

and

$$\int m_a\,da = -\lambda \sum_{b\neq a} \iint G(A, B)\,da\,db \tag{45}$$

The mass m_a, therefore, in general depends on the point A and is not constant.

The total action function in Hoyle-Narlikar's theory is then just

$$S = -\sum_a \frac{1}{2}\int m_a\,da = \lambda \sum_{a<b}\sum \iint G(A, B)\,da\,db \tag{46}$$

where $G(A, B)$ satisfies the following covariant equation:

$$\frac{\partial}{\partial x^{\mu}}\left[\sqrt{(-g)}\, g^{\mu\lambda}\frac{\partial G(x, A)}{\partial x^{\lambda}}\right]+\mu R\sqrt{(-g)}\; G(x, A) = -\delta(x, A) \qquad (47)$$

Here μ is a constant (later put $= 1/6$) and $\delta(x, A)$, the 4-dimensional δ-function. As usual, the field-equations, as well as the equation of motion are obtained from $\delta S = 0$ varying $g_{\mu\lambda}$ and the world-lines respectively.

Consider first the variation in $g_{\mu\lambda}$ such that $\delta g_{\mu\lambda} = 0$ at the boundary. The variation in the Green's function $G(A, B)$ is obtained from (47). Thus

$$\delta G(A, B) = -\int \frac{\partial}{\partial x^{\mu}}\left[\delta(\sqrt{(-g)}\, g^{\mu\lambda})\frac{\partial G(A, x)}{\partial x^{\lambda}}\right]G(B, x)\, dx$$
$$+\mu\int \delta(R\sqrt{-g})G(A, x)G(B, x)\, dx$$

which after partial integration gives

$$\delta G(A, B) = -\int \delta(\sqrt{(-g)}\, g^{\mu\lambda})\frac{\partial G(A, x)}{\partial x^{\lambda}}\frac{\partial G(B, x)}{\partial x^{\mu}}\, dx$$
$$+\mu\int \delta(R\sqrt{-g})G(A, x)G(B, x)\, dx \qquad (48)$$

The total variation of S is then

$$\delta S = \lambda\, \delta \sum_{a<b}\sum \iint G(A, B)\, da\, db$$
$$= -\sum_{a}\int m_a\, \delta(da)+\lambda\sum_{a<b}\sum\iint \delta G(A, B)\, da\, db$$

or

$$\delta S = -\sum_{a}\int m_a\, \delta(da)-\frac{1}{\lambda}\sum_{a<b}\sum\int \delta(\sqrt{(-g)}\, g^{\mu\lambda})m^{(a)}(x)_{,\lambda}m^{(b)}(x)_{,\mu}\, dx$$
$$+\frac{\mu}{\lambda}\sum_{a<b}\sum\int \delta(R\sqrt{-g})m^{(a)}(x)m^{(b)}(x)\, dx \qquad (49)$$

using the definition of the mass function according to (43). It is possible to express (49) in the usual form (cf. 1 (66))

$$\delta S = \int W_{\mu\lambda}\, \delta g^{\mu\lambda}\sqrt{(-g)}\, dx$$

with $$W_{\mu\lambda} = T_{\mu\lambda}+M_{\mu\lambda}+N_{\mu\lambda}+G_{\mu\lambda}$$

where $T_{\mu\lambda}$ coming from the first term in (49) is the familiar energy-momentum tensor of a system of particles and

$$M_{\mu\lambda} = -\frac{2\mu}{\lambda}\sum_{a<b}\sum [m^{(a)}(g_{\mu\lambda}g^{\alpha\beta}m^{(b)}_{,\alpha;\beta}-m^{(b)}_{,\alpha;\lambda})+m^{(b)}(g_{\mu\lambda}g^{\alpha\beta}m^{(a)}_{,\alpha;\beta}-m^{(a)}_{,\mu;\lambda})]$$

$$N_{\mu\lambda} = -\frac{1}{\lambda}\sum_{a<b}\sum [(1-2\mu)(m^{(a)}_{,\mu}m^{(b)}_{,\lambda}+m^{(a)}_{,\lambda}m^{(b)}_{,\mu})-(1-4\mu)g_{\mu\lambda}m^{(a),\beta}m^{(b)}_{,\beta}]$$

$$G_{\mu\lambda} = \frac{2\mu}{\lambda}(R_{\mu\lambda}-\tfrac{1}{2}g_{\mu\lambda}R)\sum_{a<b}\sum m^{(a)}m^{(b)} \tag{50}$$

$\delta S = 0$, therefore, imply the field-equations

$$W_{\mu\lambda} = 0$$

or

$$\frac{2\mu}{\lambda}(R_{\mu\lambda}-\tfrac{1}{2}g_{\mu\lambda}R)\sum_{a<b}\sum m^{(a)}m^{(b)}+T_{\mu\lambda}+M_{\mu\lambda}+N_{\mu\lambda} = 0 \tag{51}$$

The constant μ is now put equal to 1/6 so that the equation (47) for $G(x, A)$ becomes conformally invariant. The field-equations then become

$$\begin{aligned}(R_{\mu\lambda}-\tfrac{1}{2}g_{\mu\lambda}R)\sum_{a<b}\sum m^{(a)}m^{(b)} = {} & -3\lambda T_{\mu\lambda}+\sum_{a<b}\sum\{[m^{(a)}(g_{\mu\lambda}g^{\alpha\beta}m^{(b)}_{,\alpha;\beta}-m^{(b)}_{,\mu;\lambda})\\ & +m^{(b)}(g_{\mu\lambda}g^{\alpha\beta}m^{(a)}_{,\alpha;\beta}-m^{(a)}_{,\mu;\lambda})]\\ & +2[m^{(a)}_{,\mu}m^{(b)}_{,\lambda}+m^{(a)}_{,\lambda}m^{(b)}_{,\mu}-\tfrac{1}{2}g_{\mu\lambda}m^{(a),\beta}m^{(b)}_{,\beta}]\}\end{aligned} \tag{52}$$

It is now easy to show that Einstein's equations remain valid in a macroscopic smooth fluid approximation. Let

$$m(x) = \sum_a m^{(a)}(x)$$

be the total mass. Then $\sum_{a<b}\sum m^{(a)}m^{(b)} \simeq \frac{1}{2}m^2$ for a fluid containing large numbers of particles. The above field-equations reduce to the Einstein's equation

$$R_{\mu\lambda}-\tfrac{1}{2}g_{\mu\lambda}R = -8\pi G T_{\mu\lambda}$$

if we take m = constant = m_0

and

$$G = \frac{3}{4}\frac{\lambda}{m_0^2} \tag{53}$$

because all the other terms in (52) then vanish. The constant of gravitation therefore depends on the total mass m_0. It is therefore to be expected that, locally, the gravitational field is determined by cosmological factors, namely the total mass of the universe.

In order to calculate the gravitational interaction between two particles, for example, it is however necessary to solve the field-equations (52), which being more complex than Einstein's equation, present considerable difficulty. In principle, one could calculate the field $g_{\mu\lambda}$ or $G(x, y)$ from the field-equations (52), if the trajectories of the particles are given. And from $G(x, y)$ one could then determine the masses.

The equations of motion of m_a are obtained by considering the variation of its world-line $a^\mu(a)$ in (45). These turn out to be

$$\frac{d}{da}\left(m_a\frac{da^\mu}{da}\right)+m_a\Gamma^\mu_{\alpha\beta}\frac{da^\alpha}{da}\frac{da^\beta}{da}-g^{\mu\alpha}\frac{\partial m_a}{\partial a^\alpha}=0 \tag{54}$$

It is therefore necessary to specify not only the field $g_{\mu\lambda}$ at a point but also the trajectories of all *other* particles to determine the motion of m_a.

The world-lines of particles are therefore not in general geodesics. It is, however, possible to envisage a cosmological situation in which the masses are so distributed that m_a becomes constant, in which case the world-lines become geodesics.

Following the analogue of the Einstein-Infeld-Hoffmann method it may be possible to determine the trajectories as well as masses of a system of N particles from the field equations alone.

Whatever future this theory† may have it must be considered as the first theory of its kind in which cosmological factors are brought to play an essential role in determining the local behaviour of a physical system.

B. Quantum Theory

1. *Quantum theory of fields*

Free fields. We have seen in quantum mechanics that the quantization procedure imparts the characteristics of a field to a classical particle. The converse, as we shall see, also holds, namely, a classical field, after quantization, exhibits particle characteristics.‡ This enables us to describe a classical field-particle system as a quantized pure field-field system. Thus the very quantum duality enables us to remove the classical field-particle duality.

Let $\varphi_\alpha(x)$ characterize an arbitrary free field. We assume that the field-equations are derivable from a Lagrangian density

$$\mathscr{L} \equiv \mathscr{L}[\varphi_\alpha(x), \varphi_{\alpha,\mu}(x)] \tag{55}$$

† Deser, S. and Pirani, F. A. E. (1965) *Proc. R. Soc.* **A. 288**, 133; see for a criticism that it is not a pure particle theory.

‡ Cf. W. Heisenberg: "Physical Principles of Quantum Theory" Dover, New York, 1950, for a demonstration that a many-particle system is equivalent to a quantized (*second quantization*) field system.

and the variational principle

$$\delta \int \mathscr{L}\, dx = 0; \qquad dx = dx_1\, dx_2\, dx_3\, idx_0 \tag{56}$$

We use here the natural units $c = \hbar = 1$.

The field-equations are then

$$\frac{\partial}{\partial x_\mu}\left(\frac{\partial \mathscr{L}}{\partial \varphi_{\alpha,\mu}}\right) - \frac{\partial \mathscr{L}}{\partial \varphi_\alpha} = 0 \tag{57}$$

In the classical theory one then defines the canonical conjugate momentum

$$\pi_\alpha(x) = \frac{\partial \mathscr{L}}{\partial\left(\dfrac{\partial \varphi_\alpha}{\partial x_0}\right)} \tag{58}$$

and the Hamiltonian

$$H = \int \mathscr{H}\, d^3x = \int d^3x \left[\pi_\alpha(x)\frac{\partial \varphi_\alpha}{\partial x_0} - \mathscr{L}\right] \tag{59}$$

$$d^3x = dx_1\, dx_2\, dx_3$$

In quantum field theory the space-time point x—in contrast to the position operator in quantum mechanics of particles—is only a parameter and not an operator. But $\varphi_\alpha(x)$ is to be regarded as an operator attached to the point x. So that $\varphi_\alpha(x)$ and $\varphi_\alpha(x')$ at two different points are two different operators.

For a certain class of fields (Boson fields) the field operators are assumed to satisfy the following commutation relations:

$$x_0 = x'_0; \qquad \left.\begin{aligned} \left[\pi_\alpha(x), \varphi_\beta(x')\right] &= -i\, \delta_{\alpha\beta}\, \delta(\mathbf{x}-\mathbf{x}') \\ \left[\varphi_\alpha(x), \varphi_\beta(x')\right] &= \left[\pi_\alpha(x), \pi_\beta(x')\right] = 0 \end{aligned}\right\} \tag{60}$$

If the above relations are valid at any time $x_0 = x'_0$, they are also satisfied at any other equal time.

Consider an infinitesimal Lorentz transformation

$$x'_\mu = x_\mu + \varepsilon_{\mu\nu} x_\nu + \delta_\mu \tag{61}$$

Here δ_μ represents an infinitesimal translation and $\varepsilon_{\mu\nu} = -\varepsilon_{\nu\mu}$ an infinitesimal rotation. The field variable then transforms in the following form

$$\varphi'_\alpha(x') = \tfrac{1}{2}\varepsilon_{\mu\nu} S_{\mu\nu\alpha\beta}\varphi_\beta(x) + \varphi_\alpha(x) \tag{62}$$

where $S_{\mu\nu\alpha\beta}$ depends on the transformation properties of $\varphi_\alpha(x)$.

Consider now the change $\delta\varphi_\alpha(x) \equiv \varphi'_\alpha(x') - \varphi_\alpha(x')$. We have then

$$\delta\varphi_\alpha(x) = \tfrac{1}{2}\varepsilon_{\mu\nu} S_{\mu\nu\alpha\beta}\varphi_\beta(x) - \varphi_{\alpha,\mu}(x)(\varepsilon_{\mu\nu}x_\nu + \delta_\mu) \tag{63}$$

The change in $\pi_\alpha(x)$ can be similarly calculated

$$\delta\pi_\alpha(x) = -\tfrac{1}{2}\varepsilon_{\lambda\nu}S_{\lambda\nu\beta\alpha}\pi_\beta(x) - \pi_{\alpha,\lambda}(\varepsilon_{\lambda\nu}x_\nu + \delta_\lambda) - i\varepsilon_{\alpha\nu}\pi_{\alpha\nu} \tag{64}$$

where $\pi_{\alpha\nu} \equiv \partial\mathscr{L}/\partial\varphi_{\alpha,\nu}$. To ensure Lorentz invariance of our theory we must now demand that $\varphi_\alpha(x)+\delta\varphi_\alpha(x)$ and $\pi_\alpha(x)+\delta\pi_\alpha(x)$ satisfy the same commutation relations as $\varphi_\alpha(x)$ and $\pi_\alpha(x)$. This implies that there exists a Hermetian operator T such that

$$\left.\begin{aligned} \varphi_\alpha(x)+\delta\varphi_\alpha(x) &= e^{iT}\varphi_\alpha(x)e^{-iT} \\ \pi_\alpha(x)+\delta\pi_\alpha(x) &= e^{iT}\pi_\alpha(x)e^{-iT} \end{aligned}\right\} \tag{65}$$

Since we are considering infinitesimal transformations we can write instead of (65)

$$\left.\begin{aligned} \delta\varphi_\alpha(x) &= i[T, \varphi_\alpha(x)] \\ \delta\pi_\alpha(x) &= i[T, \pi_\alpha(x)] \end{aligned}\right\} \tag{66}$$

We give T explicitly:

$$T = \int d^3x\big[\tfrac{1}{2}\varepsilon_{\mu\nu}S_{\mu\nu\alpha\beta}\pi_\alpha(x)\varphi_\beta(x) + x_\nu\varepsilon_{\nu k}\pi_\alpha(x)\varphi_{\alpha,k}(x) + i(\varepsilon_{4k}x_k+\delta_4)\Big(\pi_\alpha(x)\frac{\partial\varphi_\alpha(x)}{\partial x_0} - \mathscr{L}\Big) - \delta_k\pi_\alpha(x)\varphi_{\alpha,k}(x)\big]$$

It can also be expressed as

$$T = -i\int d^3x\, T_{4\nu}\,\delta x_\nu \tag{67}$$

where

$$\begin{aligned} T_{\mu\nu} &= -\pi_{\alpha\mu}(x)\varphi_{\alpha,\nu}(x) + \delta_{\mu\nu}\mathscr{L} - f_{\lambda\mu\nu,\lambda} \\ f_{\lambda\mu\nu} &= \tfrac{1}{2}[S_{\mu\nu\alpha\beta}\pi_{\alpha\lambda}(x)\varphi_\beta(x) + S_{\nu\lambda\alpha\beta}\pi_{\alpha\mu}(x)\varphi_\beta(x) + S_{\mu\lambda\alpha\beta}\pi_{\alpha\nu}(x)\varphi_\beta(x)] \\ \delta x_\nu &= \varepsilon_{\nu\lambda}x_\lambda + \delta_\nu \end{aligned}$$

From the field-equations (57) it follows that

$$T_{\mu\nu,\mu} = 0 \tag{68}$$

so that $P_\mu = -i\int d^3x\, T_{4\mu}$ is a constant of motion. We have

$$\left.\begin{aligned} P_k &= -\int d^3x\, \pi_\alpha(x)\varphi_{\alpha,k}(x) \\ P_4 &= iH \end{aligned}\right\} \tag{69}$$

If we set $\varepsilon_{\nu\lambda} = 0$, that is, consider only translations, we get

$$\left.\begin{aligned} \delta\varphi_\alpha(x) &= -\varphi_{\alpha,\mu}\,\delta_\mu \\ \delta\pi_\alpha(x) &= -\pi_{\alpha,\mu}\,\delta_\mu \\ T_{\text{tran}} &= P_\mu\,\delta_\mu \end{aligned}\right\} \tag{70}$$

So that from (66) and (70), since δ_μ is arbitrary

$$\left.\begin{aligned} i\varphi_{\alpha,\mu}(x) &= [P_\mu, \varphi_\alpha(x)] \\ i\pi_{\alpha,\mu}(x) &= [P_\mu, \pi_\alpha(x)] \end{aligned}\right\} \tag{71}$$

P_μ is called the energy-momentum operator of the field system. Its existence follows from Lorentz-invariance of the theory. The corresponding operator for rotation is given by

$$\begin{aligned} T_{\text{rot}} &= -i\int d^3x\, T_{4\nu}\varepsilon_{\nu\lambda}x_\lambda \\ &= -\tfrac{1}{2}\varepsilon_{\nu\lambda}J_{\nu\lambda} \end{aligned}$$

where

$$J_{\nu\lambda} = i\int d^3x[T_{4\nu}x_\lambda - T_{4\lambda}x_\nu] \tag{72}$$

is called the angular momentum operator of the field system.

A classic example of Boson fields is the free electromagnetic field characterized by the vector potential $A_\mu(x)$. Its field-equations can be derived from the Lagrangian density

$$\mathscr{L} \equiv -\tfrac{1}{4}f_{\mu\lambda}f_{\mu\lambda} - \tfrac{1}{2}A_{\mu,\mu}A_{\nu,\nu} \tag{73}$$

$$f_{\mu\nu} = A_{\nu,\mu} - A_{\mu,\nu}$$

The field-equations are then

$$\Box\, A_\mu = 0 \tag{74}$$

In order that the above be identical with the Maxwell's equations for $f_{\mu\lambda}$ it is sufficient to demand the subsidiary condition

$$\left.\begin{aligned} A_{\nu,\nu} &= 0 \\ \frac{\partial}{\partial x_0}A_{\nu,\nu} &= 0 \end{aligned}\right\} \text{for all } \mathbf{x} \tag{75}$$

The canonical conjugate momenta are then

$$\left.\begin{aligned} \pi_k &= if_{4k} \\ \pi_4 &= iA_{\nu,\nu} \end{aligned}\right\} \tag{76}$$

In the formal quantization of the electromagnetic field the subsidiary conditions (75) cannot be carried over as they now stand. We shall come back to them later.

The commutation relations (60) are equivalent to

$$\begin{gathered} [A_\mu(x), A_\nu(x')] = 0 \\ \left[\frac{\partial A_\mu(x)}{\partial x_0}, A_\nu(x')\right] = -i\,\delta_{\mu\nu}\,\delta(\mathbf{x}-\mathbf{x}') \\ \left[\frac{\partial A_\mu(x)}{\partial x_0}, \frac{\partial A_\nu(x')}{\partial x_0'}\right] = 0 \end{gathered} \tag{77}$$

at $x_0 = x_0'$.

We shall frequently consider our field system to be confined in a box of volume V and impose periodic boundary conditions. We can then expand $A_\mu(x)$ in its Fourrier components

$$A_\mu(x) = \frac{1}{\sqrt{V}} \sum e^{ikx} A_\mu(k) \tag{78}$$

$$kx = \sum_\mu k_\mu x_\mu$$

In view of (74) we must have $k^2 = \mathbf{k}^2 - k_0^2 = 0$

or $\qquad k_0 = \pm\omega; \qquad \omega = +\sqrt{\mathbf{k}^2}$

so that (78) becomes

$$A_\mu(x) = \frac{1}{\sqrt{V}} \sum_{\mathbf{k}} [e^{ikx} A_\mu(\mathbf{k}) + e^{-ikx} A_\mu^*(\mathbf{k})] \tag{79}$$

$$k_0 = \omega$$

The reality conditions on $A_\mu(x)$ imply that $A_k(\mathbf{k})$, $iA_4(\mathbf{k})$ and $A_k^*(\mathbf{k})$, $iA_4^*(\mathbf{k})$ are to be Hermetian conjugate to each other. Furthermore, since $A_\mu(x)$ is a vector, for every $\mathbf{k}$ there exist four independent polarized components. The general solution of (74) can therefore be written as

$$A_\mu(x) = \frac{1}{\sqrt{V}} \sum_{\mathbf{k}} \sum_{\lambda=1}^{4} \frac{e_\mu^{(\lambda)}}{\sqrt{(2\omega)}} [e^{ikx} a^{(\lambda)}(\mathbf{k}) + e^{-ikx} a^{*(\lambda)}(\mathbf{k})] \tag{80}$$

$$e_\mu^{(\lambda)} e_\mu^{(\lambda')} = \delta_{\lambda\lambda'}$$

Here $e_\mu^{(\lambda)}$ are four polarization vectors which may be taken as follows

$$\begin{aligned} e_4^{(k)} &= 0 \\ e_l^{(1)}k_l &= e_l^{(2)}k_l = 0 \\ e_l^{(3)} &= \frac{k_l}{\omega} \\ e_l^{(4)} &= 0; \qquad e_4^{(4)} = 1 \end{aligned} \tag{81}$$

so that we also have

$$\sum_\lambda e_\mu^{(\lambda)} e_\nu^{(\lambda)} = \delta_{\mu\nu}$$

The factor $1/\sqrt{(2\omega)}$ is introduced in order that the commutation relations for $a(\mathbf{k})$ take the simple form (in view of (77))

$$\begin{aligned} \left[a^{(\lambda)}(\mathbf{k}), a^{(\lambda')}(\mathbf{k}')\right] &= \left[a^{*(\lambda)}(\mathbf{k}), a^{*(\lambda')}(\mathbf{k}')\right] = 0 \\ \left[a^{(\lambda)}(\mathbf{k}), a^{*(\lambda')}(\mathbf{k}')\right] &= \delta_{\lambda\lambda'}\, \delta_{\mathbf{k}\mathbf{k}'} \end{aligned} \tag{82}$$

From (59) and (73) we obtain for the Hamiltonian operator

$$\begin{aligned} H &= \tfrac{1}{2} \int d^3x \left[\frac{\partial A_\mu}{\partial x_0}\frac{\partial A_\mu}{\partial x_0} + A_{\mu,k}A_{\mu,k}\right] \\ &= \tfrac{1}{2} \sum_{\mathbf{k},\lambda} \omega\{a^{(\lambda)}(\mathbf{k}), a^{*(\lambda)}(\mathbf{k})\} \end{aligned} \tag{83}$$

where $\quad \{a^{(\lambda)}(\mathbf{k}), a^{*(\lambda)}(\mathbf{k})\} = a^{(\lambda)}(\mathbf{k})a^{*(\lambda)}(\mathbf{k}) + a^{*(\lambda)}(\mathbf{k})a^{(\lambda)}(\mathbf{k})$

Similarly the momentum operator P_k becomes

$$\begin{aligned} P_k &= -\int d^3x[i(A_{l,4} - A_{4,l})A_{l,k} + iA_{\nu,\nu}A_{4,k}] \\ &= \tfrac{1}{2} \sum_{\mathbf{k},\lambda} k_k\{a^{(\lambda)}(\mathbf{k}), a^{*(\lambda)}(\mathbf{k})\} \end{aligned} \tag{84}$$

If we now introduce the operators

$$\begin{aligned} N^{(\lambda)}(\mathbf{k}) &= a^{*(\lambda)}(\mathbf{k})a^{(\lambda)}(\mathbf{k}), \qquad \lambda \neq 4 \\ N^{(4)}(\mathbf{k}) &= -a^{(4)}(\mathbf{k})a^{*(4)}(\mathbf{k}) \end{aligned} \tag{85}$$

we can also write in view of (82)

$$\begin{aligned} P_k &= \sum_{\mathbf{k}} k_k\left(\sum_{\lambda=1}^{3} N^{(\lambda)}(\mathbf{k}) - N^{(4)}(\mathbf{k})\right) \\ H &= \sum_{\mathbf{k}} \omega\left(\sum_{\lambda=1}^{3} N^{(\lambda)}(\mathbf{k}) - N^{(4)}(\mathbf{k})\right) \end{aligned} \tag{86}$$

where we have left out the so-called zero-point contributions. From the commutation relations (82) it follows that the eigenvalues of $N^{(\lambda)}(\mathbf{k})$ are positive integers. Our field system therefore behaves as an assembly of particles (photons) of energy ω and momentum $\mathbf{k}$. The operators $a^{(\lambda)}(\mathbf{k})$ for $\lambda \neq 4$ and $-a^{*(4)}(\mathbf{k})$ can be regarded as "destruction" operators and $a^{*(\lambda)}(\mathbf{k})$ for $\lambda \neq 4$ and $a^{(4)}(\mathbf{k})$ as "creation" operators.

The above expression for the energy as it stands is not positive definite. We now consider the subsidiary condition (75) which as mentioned earlier cannot be considered as a formal operator relation. In order to ensure that the quantized theory should lead to the classical Maxwell theory in the limiting case it is not *necessary* to demand that all classical relations be carried over as formal operator relations. It is *sufficient* to demand that the *expectation* values of the field variables in any physically realizable state should satisfy the classical relations. It is therefore sufficient to restrict the physically realizable states $|\psi\rangle$ by the new subsidiary condition

$$A_{\nu,\nu}|\psi\rangle = |0\rangle \quad \text{(null vector)} \tag{87}$$

Equation (87) is equivalent to

$$[a^{(3)}(\mathbf{k})+ia^{(4)}(\mathbf{k})]|\psi\rangle = |0\rangle \tag{88}$$

$$[a^{*(3)}(\mathbf{k})+ia^{*(4)}(\mathbf{k})]|\psi\rangle = 0\rangle$$

Equation (88), in view of our choice of the polarization vectors, is a condition on the longitudinal and scalar photons. By a proper choice of gauge transformation these degrees of freedom can be eliminated; the total energy of the system is then due to only the transverse photons as the longitudinal and scalar photons can be made to compensate each other in view of (88). A similar analysis holds for the momentum P_k.

We are now in a position to compute the commutation relations for $A_\mu(x)$ for any arbitrary points x and x'. We note that (from (80) and (82))

$$[A_\mu(x), A_\nu(x')] = \frac{1}{V}\sum_{\mathbf{k},\mathbf{k}',\lambda,\lambda'} \frac{e_\mu{}^{(\lambda)}e^{\nu(\lambda')}}{2\sqrt{(\omega\omega')}}$$

$$\times [e^{ikx}a^{(\lambda)}(\mathbf{k})+e^{-ikx}a^{*(\lambda)}(\mathbf{k}), e^{ik'x'}a^{(\lambda')}(\mathbf{k}')+e^{-ik'x'}a^{*(\lambda')}(\mathbf{k}')]$$

$$= \frac{1}{V}\sum_{\mathbf{k},\lambda} \frac{e_\mu^{(\lambda)}e_\nu^{(\lambda)}}{2\omega}(e^{ik(x-x')}-e^{-ik(x-x')})$$

If we now let the dimensions of the volume V tend to infinity we have

$$\frac{1}{V}\sum_{\mathbf{k}} f(\mathbf{k}) \rightarrow \frac{1}{(2\pi)^3}\int f(\mathbf{k})\, d^3k$$

so that

$$[A_\mu(x), A_\nu(x')] = -i\,\delta_{\mu\nu} D(x'-x) \tag{89}$$

where
$$D(x) = -\frac{i}{(2\pi)^3}\int \frac{d^3k}{2\omega}(e^{ikx} - e^{-ikx}) \tag{90}$$

$D(x)$ can also be expressed as

$$D(x) = -(i/(2\pi)^3)\int dk e^{ikx}\,\delta(k^2)\varepsilon(k) \tag{91}$$

where $\varepsilon(k) = k_0/|k_0|$. The function $D(x)$ has the following properties:

$$\left.\begin{aligned} D(x) &= 0 \\ \frac{\partial D(x)}{\partial x_0} &= -\delta(\mathbf{x}) \end{aligned}\right\} \text{ for } x_0 = 0 \tag{92}$$

$D(x)$ vanishes not only when $x_0 = 0$ but also when $x^2 > 0$. According to (89) the commutator of $A_\mu(x)$ and $A_\nu(x')$ vanishes whenever the two points x and x' are separated by a space-like interval. Physically, this means that measurements of the field at two points separated by space-like intervals do not influence each other.

We define some more singular functions for future reference.

$$D^{(1)}(x) = \frac{1}{(2\pi)^3}\int dk e^{ikx}\,\delta(k^2) \tag{93}$$

$$D_R(x) = \begin{cases} -D(x) & \text{for } x_0 > 0 \\ 0 & \text{for } x_0 < 0 \end{cases} \tag{94}$$

$$D_A(x) = \begin{cases} 0 & \text{for } x_0 > 0 \\ D(x) & \text{for } x_0 < 0 \end{cases} \tag{95}$$

$$\bar{D}(x) = \tfrac{1}{2}[D_R(x) + D_A(x)] = -\tfrac{1}{2}\varepsilon(x)D(x) \tag{96}$$

$$D^{(+)}(x) = \tfrac{1}{2}[D(x) - iD^{(1)}(x)] \tag{97}$$

$$D^{(-)}(x) = \tfrac{1}{2}[D(x) + iD^{(1)}(x)] \tag{98}$$

$$D_F(x) = (2/i)\bar{D}(x) + D^{(1)}(x) \tag{99}$$

We have for example

$$\Box\, D(x) = 0 \tag{100}$$

$$\Box\, \bar{D}(x) = \Box\, D_R(x) = \Box\, D_A(x) = -\delta(x) \tag{101}$$

$$\Box\, D_F(x) = 2i\,\delta(x) \tag{102}$$

We consider next the free electron (positron) field characterized by a four-component spinor field $\psi \equiv \psi_\alpha(x)$ satisfying the Dirac equation

$$(\gamma_\mu \partial_\mu + m)\psi = 0 \tag{103}$$

Here m is the mass of electron (positron) and the γ_μ's are 4×4 matrices with the property

$$\{\gamma_\mu, \gamma_\lambda\} = \delta_{\mu\lambda} \tag{104}$$

An explicit representation of the γ_μ's is as follows:

$$\gamma_k = \begin{pmatrix} 0 & -i\sigma_k \\ i\sigma_k & 0 \end{pmatrix}, \qquad \gamma_4 = \begin{pmatrix} I & 0 \\ 0 & -I \end{pmatrix} \tag{105}$$

where

$$\sigma_1 = \begin{pmatrix} 0 & 1 \\ 1 & 0 \end{pmatrix}; \quad \sigma_2 = \begin{pmatrix} 0 & -i \\ i & 0 \end{pmatrix}; \quad \sigma_3 = \begin{pmatrix} 1 & 0 \\ 0 & -1 \end{pmatrix}; \quad I = \begin{pmatrix} 1 & 0 \\ 0 & 1 \end{pmatrix}$$

In this representation for every plane-wave solutions of (103)

$$\psi_\alpha(x) = u_\alpha(\mathbf{q}) e^{i(\mathbf{q}\cdot\mathbf{x} - q_0 x_0)} \tag{106}$$

for a given $\mathbf{q}$ there are two possible values of q_0

$$q_0 = \pm E = \pm\sqrt{(q^2 + m^2)} \tag{107}$$

and for every q_0 there exist two independent solutions $u_\alpha^{(r)}(\mathbf{q})$ which have the following property

$$\sum_\alpha u_\alpha^{*(r)} u_\alpha^{(s)} = \delta_{rs} \tag{108}$$

$$\sum_r u_\alpha^{*(r)} u_\beta^{(r)} = \delta_{\alpha\beta} \tag{109}$$

Here the star represents the complex conjugate. Equations (108) and (109) are, respectively, the orthogonality and completeness relations for the $u_\alpha^{(r)}(\mathbf{q})$.

The Dirac equation can be derived from the following Lagrangian density:

$$\mathscr{L} = -\bar{\psi}(x)(\gamma_\mu \partial_\mu + m)\psi(x) \tag{110}$$

Here $\bar{\psi} = \psi^* \gamma_4$; ψ^* is the adjoint of the column matrix ψ.

The Hamiltonian density can be written as

$$\mathscr{H} = -i\pi_\psi(x)\gamma_4(\gamma_k \partial_k + m)\psi(x) \tag{111}$$

where $$\pi_\psi(x) = i\bar{\psi}(x)\gamma_4 = i\psi^*(x)$$

One defines the charge-current density $j_\mu(x)$ by

$$j_\mu(x) = ie\bar{\psi}(x)\gamma_\mu\psi(x) \tag{112}$$

$j_\mu(x)$ satisfies the continuity equation $j_{\mu,\mu} = 0$.

As in the electromagnetic case we now expand the electron-positron field in terms of the plane wave solutions (106).

$$\psi_\alpha(x) = \frac{1}{\sqrt{V}}\sum_{\mathbf{q}}\left\{e^{i(\mathbf{q}\cdot\mathbf{x}-Ex_0)}\sum_{r=1}^{2} u_\alpha^{(r)}(\mathbf{q})a^{(r)}(\mathbf{q})+e^{i(\mathbf{q}\cdot\mathbf{x}+Ex_0)}\sum_{r=3}^{4} u_\alpha^{(r)}(\mathbf{q})a^{(r)}(\mathbf{q})\right\} \tag{113}$$

The variables $a^{(r)}(\mathbf{q})$ are now operators in the quantized version. The Hamiltonian operator then becomes

$$H = \int d^3x\, \mathscr{H} = \sum_{\mathbf{q}}\left[E\sum_{r=1}^{2} a^{*(r)}(\mathbf{q})a^{(r)}(\mathbf{q})-\sum_{r=3}^{4} a^{*(r)}(\mathbf{q})a^{(r)}(\mathbf{q})\right]$$

and the charge

$$Q = -i\int j_4\, d^3x = e\sum_{\mathbf{q}}\sum_{r=1}^{4} a^{*(r)}(\mathbf{q})a^{(r)}(\mathbf{q})$$

The commutation relations for the basic operators, in contrast to the electromagnetic case, are now assumed to be

$$\{a^{*(r)}(\mathbf{q}),\, a^{(s)}(\mathbf{q}')\} = \delta_{rs}\,\delta_{\mathbf{q}\mathbf{q}'} \tag{114}$$

$$\{a^{*(r)}(\mathbf{q}),\, a^{*(s)}(\mathbf{q}')\} = \{a^{(r)}(\mathbf{q}),\, a^{(s)}(\mathbf{q}')\} = 0 \tag{115}$$

We can then write for the Hamiltonian

$$H = \sum_{\mathbf{q}} E\sum_{r=1}^{2} (N^{+(r)}(\mathbf{q})+N^{-(r)}(\mathbf{q})) \tag{116}$$

where

$$N^{+(r)}(\mathbf{q}) = a^{*(r)}(\mathbf{q})a^{(r)}(\mathbf{q})$$

$$N^{-(r)}(\mathbf{q}) = b^{*(r)}(\mathbf{q})b^{(r)}(\mathbf{q})$$

$$b^{(1)}(\mathbf{q}) = a^{*(4)}(-\mathbf{q})$$

$$b^{(2)}(\mathbf{q}) = a^{*(3)}(-\mathbf{q})$$

In (116) we have left out the so-called zero-point energy. The corresponding expression for the charge without the zero-point term is

$$Q = e\sum_{\mathbf{q},r} [N^{+(r)}(\mathbf{q})-N^{-(r)}(\mathbf{q})] \tag{117}$$

We can therefore regard N^+ and N^- as the particle number operators for electrons and positrons respectively.

The zero-point term in the original expression for Q can be eliminated if we redefine the charge-current density by

$$j_\mu = (ie/2)[\bar{\psi},\, \gamma_\mu\psi] \tag{118}$$

j_μ then possesses the following charge-symmetry, namely it is invariant under the transformation

$$N^+ \rightleftarrows N^-$$

$$e \leftarrow -e$$

We can make the theory charge-symmetric from the very beginning by taking the following symmetric Lagrangian density

$$\mathscr{L} = -\tfrac{1}{4}[\bar{\psi}, (\gamma_\mu \partial_\mu + m)\psi] - \tfrac{1}{4}[(-\partial_\mu \bar{\psi}\gamma_\mu + m\bar{\psi}), \psi] \tag{119}$$

The commutation relations for the field operators now become

$$\begin{aligned} \{\bar{\psi}_\alpha(x), \psi_\beta(x')\} &= -iS_{\beta\alpha}(x'-x) \\ \{\psi(x), \psi(x')\} &= \{\bar{\psi}(x), \bar{\psi}(x')\} = 0 \end{aligned} \tag{120}$$

with

$$S_{\alpha\beta}(x) = -(i/(2\pi)^3) \int dq\, e^{iqx}(i\gamma q - m)_{\alpha\beta}\, \delta(q^2+m^2)\varepsilon(q) \tag{121}$$

where $\gamma q = \sum_\mu \gamma_\mu q_\mu$. It is now possible to define the corresponding functions S_R, S_A, $\bar{S}$, S_F according to (94)–(99), starting from

$$S^{(1)}(x) = (1/(2\pi)^3) \int dq\, e^{iqx}(i\gamma q - m)\, \delta(q^2+m^2) \tag{122}$$

Interacting fields. One obtains a completely field theoretic description of the phenomena of electrodynamics by coupling the free electromagnetic and the electron (positron) field with an interaction term in the total Lagrangian $\mathscr{L}$.

$$\mathscr{L} = \mathscr{L}_\psi + \mathscr{L}_A + \mathscr{L}_I \tag{123}$$

$$\mathscr{L}_\psi = -\tfrac{1}{4}[\bar{\psi}, (\gamma_\mu \partial_\mu + m)\psi] - \tfrac{1}{4}[(-\partial_\mu \bar{\psi}\gamma_\mu + m\bar{\psi}), \psi]$$

$$\mathscr{L}_A = -\tfrac{1}{4} f_{\mu\nu} f_{\mu\nu} - \tfrac{1}{2} A_{\mu,\mu} A_{\nu,\nu}$$

$$\mathscr{L}_I = A_\mu j_\mu; \qquad j_\mu = (ie/2)[\bar{\psi}, \gamma_\mu \psi]$$

The field-equations of electrodynamics then become

$$\begin{aligned} (\gamma_\mu \partial_\mu + m)\psi &= ie\gamma_\mu A_\mu \psi \\ \Box\, A_\mu &= -j_\mu = -(ie/2)[\bar{\psi}, \gamma_\mu \psi] \end{aligned} \tag{124}$$

The interaction term does not contain any time derivatives of the field operators; so that the canonical operators π_ψ and π_A are same functions of the field operators as in the free-field case. Even in the presence of interaction the field operators satisfy therefore the same commutation relations (89) and

(120) *at* $x_0 = x_0'$. To obtain the commutation relations at $x_0 \neq x_0'$ it is, however, necessary to solve the field equations (124).

In physics one is interested in the following problem. Given initially the number and energies of incoming photons and electrons (positrons) what is the probability of finding a given number of outgoing photons and electrons (positrons) of given energies after a short period of interaction. Let $\psi^{(\text{in})}$, $A_\mu^{(\text{in})}$ and $\psi^{(\text{out})}$, $A_\mu^{(\text{out})}$ denote the incoming and outgoing fields respectively. Instead of the differential equations (124) we can write the following integral expressions:

$$\begin{aligned}\psi(x) &= \psi^{(\text{in})}(x) - \int S_R(x-x')ie\gamma_\mu A_\mu(x')\psi(x')\,dx' \\ &= \psi^{(\text{out})}(x) - \int S_A(x-x')ie\gamma_\mu A_\mu(x')\psi(x')\,dx' \end{aligned} \tag{125}$$

$$\begin{aligned}A_\mu(x) &= A_\mu^{(\text{in})}(x) + \int D_R(x-x')j_\mu(x')\,dx' \\ &= A_\mu^{(\text{out})}(x) + \int D_A(x-x')j_\mu(x')\,dx' \end{aligned} \tag{126}$$

where the incoming and outgoing fields satisfy the free-field equations

$$\begin{aligned}(\gamma_\mu\partial_\mu + m)\psi^{(\text{in})}(x) &= (\gamma_\mu\partial_\mu + m)\psi^{(\text{out})}(x) = 0 \\ \Box\, A_\mu^{(\text{in})}(x) &= \Box\, A_\mu^{(\text{out})}(x) = 0\end{aligned} \tag{127}$$

It follows from the definition of the so-called retarded and advanced functions S_R, D_R and S_A, D_A, that

$$\left.\begin{aligned}\psi(x) &\to \psi^{(\text{in})}(x) \\ A_\mu(x) &\to A_\mu^{(\text{in})}(x)\end{aligned}\right\} \quad \text{for } x_0 \to -\infty$$

$$\left.\begin{aligned}\psi(x) &\to \psi^{(\text{out})}(x) \\ A_\mu(x) &\to A_\mu^{(\text{out})}(x)\end{aligned}\right\} \quad \text{for } x_0 \to +\infty$$

Also the total Hamiltonian $H(\psi, A) \to H^0(\psi^{(\text{in})}, A^{(\text{in})})$ for $x_0 \to -\infty$ and $H(\psi, A) \to H^0(\psi^{(\text{out})}, A^{(\text{out})})$ for $x_0 \to +\infty$, where H^0 is the sum of the Hamiltonians of the *free* electromagnetic and electron (positron) fields. Since the incoming and outgoing fields satisfy the free-field equations they also obey the same commutation relations (89) and (120). It follows therefore that there exists an operator S such that

$$\begin{aligned}\psi^{(\text{out})} &= S^{-1}\psi^{(\text{in})}S \\ A_\mu^{(\text{out})} &= S^{-1}A_\mu^{(\text{in})}S \\ SS^* &= S^*S = I\end{aligned} \tag{128}$$

We also have for example

$$H^0(\psi^{(\text{out})}, A^{(\text{out})}) = S^{-1}H^0(\psi^{(\text{in})}, A^{(\text{in})})S \tag{129}$$

Let $|n\rangle$ be an eigenstate of $H^0(\psi^{(\text{in})}, A^{(\text{in})})$ with energy E_n, i.e.,

$$H^0(\psi^{(\text{in})}, A^{(\text{in})})|n\rangle = E_n|n\rangle$$

From (129) we have

$$H^0(\psi^{(\text{out})}, A^{(\text{out})})S^{-1}|n\rangle = E_nS^{-1}|n\rangle$$

so that $S^{-1}|n\rangle$ is an eigenstate of the energy for $x_0 \to +\infty$. The transition probability $\omega_{nn'}$ that a state $|n\rangle$ makes a transition to a state $|n'\rangle$ due to the interaction is therefore given by

$$\omega_{nn'} = |\langle n'|S|n\rangle|^2 \tag{130}$$

To calculate S we have in principle the following equations:
(cf. (125) and (126))

$$\psi^{(\text{out})}(x) = S^{-1}\psi^{(\text{in})}(x)S = \psi^{(\text{in})}(x) + \int S(x-x')ie\gamma_\mu A_\mu(x')\psi(x')\,dx'$$

$$A_\mu^{(\text{out})}(x) = S^{-1}A_\mu^{(\text{in})}(x)S = A_\mu^{(\text{in})}(x) - \int D(x-x')j_\mu(x')\,dx'$$

or

$$\begin{aligned}[S, \psi^{(\text{in})}(x)] &= -S\int S(x-x')ie\gamma_\mu A_\mu(x')\psi(x')\,dx' \\ [S, A_\mu^{(\text{in})}(x)] &= S\int D(x-x')j_\mu(x')\,dx'\end{aligned} \tag{131}$$

A more effective method is to introduce the so-called "interaction picture". We have seen that at any given time x_0 the field operators $\psi(x)$ and $A_\mu(x)$ satisfy the same commutation relations as the incoming or outgoing operators. Therefore, there exists a time-dependent operator $U(x_0)$ such that

$$\begin{aligned}\psi(x) &= U^{-1}(x_0)\psi^{(\text{in})}(x)U(x_0) \\ A_\mu(x) &= U^{-1}(x_0)A_\mu^{(\text{in})}(x)U(x_0) \\ U^*(x_0)U(x_0) &= U(x_0)U^*(x_0) = I\end{aligned} \tag{132}$$

$U(x_0)$ then satisfies the following differential equation

$$i\frac{\partial U(x_0)}{\partial x_0} = H^{\text{int}}(A_\mu^{(\text{in})}, \psi^{(\text{in})})U(x_0) \tag{133}$$

with the initial condition $U(-\infty) = I$. Here $H^{(\text{int})}$ is the interaction term in the total Hamiltonian H. The S operator is then given by $U(+\infty) = S$. We can write (133) in the following integral form

$$U(x_0) = I - i \int_{-\infty}^{x_0} dx_0' \, H^{(\text{int})}(A^{(\text{in})}(x'), \psi^{(\text{in})}(x')) U(x_0') \tag{134}$$

We treat $H^{(\text{int})}$ as a perturbation and obtain a series solution

$$U(x_0) = \sum_{n=0}^{\infty} (-i)^n \int_{-\infty}^{x_0} dx_0' \int_{-\infty}^{x_0'} dx_0'' \cdots \int_{-\infty}^{x_0^{(n-1)}} dx_0^{(n)} H^{(\text{int})}(x_0') \cdots H^{(\text{int})}(x_0^{(n)}) \tag{135}$$

If one introduces the following chronological operator:

$$P(A(x_0)B(x_0')) = \begin{cases} A(x_0)B(x_0') & \text{for } x_0 > x_0' \\ B(x_0')A(x_0) & \text{for } x_0' > x_0 \end{cases} \tag{136}$$

one can transform the integral in (135) into a symmetric expression, so that we have for the S-operator

$$S = I + \sum_{n=1}^{\infty} (-i)^n/n! \int_{-\infty}^{\infty} dx_0' \cdots \int_{-\infty}^{\infty} dx_0^{(n)} \, P(H^{(\text{int})}(x') \cdots H^{(\text{int})}(x_0^{(n)})) \tag{137}$$

Let the initial and final states be characterized by

$$|n\rangle = |n_k\rangle |n_p\rangle \quad \text{and} \quad |n'\rangle = |n_k'\rangle |n_p'\rangle$$

where $|n_k\rangle$ and $|n_p\rangle$ are eigenstates of the Hamiltonians of free electromagnetic and electron (positron) fields, respectively. The interaction Hamiltonian $H^{(\text{int})}$ is simply $\int [\overline{\psi}^{(\text{in})}, \gamma_\mu \psi^{(\text{in})}] A_\mu^{(\text{in})} \, d^3x$. The total S-matrix is then the infinite sum of $\langle n'|S^{(n)}|n\rangle$ where

$$\begin{aligned} \langle n'|S^{(n)}|n\rangle = \frac{(-e)^n}{n!2^n} \int dx' \cdots \int dx^{(n)} \langle n_q'| P([\overline{\psi}^{(\text{in})}(x'), \gamma_{\nu_1}\psi^{(\text{in})}(x')] \\ \cdots [\overline{\psi}^{(\text{in})}(x^{(n)}), \gamma_{\nu_n}\psi^{(\text{in})}(x^{(n)})]) |n_q\rangle \\ \times \langle n_k'| P(A_{\nu_1}^{(\text{in})}(x') \cdots A_{\nu_n}^{(\text{in})}(x^{(n)})) |n_k\rangle \end{aligned} \tag{138}$$

Here the integration is over the entire space-time; it is possible to separate the electromagnetic field operators from the electron (positron) ones as they commute with each other. The above expression can be further simplified as follows.

We have for example

$$A_\mu^{(\text{in})}(x) = A_\mu^{(+)}(x) + A_\mu^{(-)}(x) \tag{139}$$

where

$$A_\mu^{(+)}(x) = \frac{1}{\sqrt{V}} \sum_{\mathbf{k},\lambda} \frac{e^{(\lambda)}}{\sqrt{(2\omega)}} e^{ikx} a^{(\lambda)}(\mathbf{k})$$

$$A_\mu^{(-)}(x) = \frac{1}{\sqrt{V}} \sum_{\mathbf{k},\lambda} \frac{e^{(\lambda)}}{\sqrt{(2\omega)}} e^{-ikx} a^{*(\lambda)}(\mathbf{k})$$

$$[A_\mu^{(+)}(x), A_\nu^{(+)}(x')] = [A_\mu^{(-)}(x), A_\nu^{(-)}(x')] = 0$$

and (140)

$$[A_\mu^{(+)}(x), A_\nu^{(-)}(x')] = -i\,\delta_{\mu\nu} D^{(-)}(x-x') \equiv \langle 0'|A_\mu^{(\mathrm{in})}(x) A_\nu^{(\mathrm{in})}(x')|0'\rangle$$

On the right-hand side we have an expectation value on the vacuum state† $|0'\rangle$. Similarly

$$\psi^{(\mathrm{in})}(x) = \psi^{(+)}(x) + \psi^{(-)}(x);\ \bar{\psi}^{(\mathrm{in})}(x) = \bar{\psi}^{(+)}(x) + \bar{\psi}^{(-)}(x)$$

$$\{\psi^{(+)}(x), \psi^{(+)}(x')\} = \{\bar{\psi}^{(+)}(x), \psi^{(+)}(x')\} = \{\psi^{(+)}(x), \psi^{(-)}(x')\} \ldots = 0$$

except that

$$\begin{aligned} \{\bar{\psi}^{(+)}(x), \psi^{(-)}(x')\} &= -iS^{(-)}(x'-x) \equiv \langle 0'|\bar{\psi}^{(\mathrm{in})}(x)\psi^{(\mathrm{in})}(x')|0'\rangle \\ \{\bar{\psi}^{(-)}(x), \psi^{(+)}(x')\} &= -iS^{(+)}(x'-x) \equiv \langle 0'|\psi^{(\mathrm{in})}(x)\bar{\psi}^{(\mathrm{in})}(x')|0'\rangle \end{aligned} \quad (141)$$

Let us denote

$$A(i) = A_{\mu_i}^{(\mathrm{in})}(x^{(i)})$$

$$\varphi(i) = \psi^{(\mathrm{in})}(x^{(i)}) \quad \text{or} \quad \bar{\psi}^{(\mathrm{in})}(x^{(i)})$$

Define now the normal product of n operators by

$$\begin{aligned} N(A(1)\ldots A(n)) = {} & A^{(+)}(1)\ldots A^{(+)}(n) + \sum_{i=1}^{n} A^{(-)}(i) A^{(+)}(1)\ldots A^{(+)}(n) \\ & + \sum_{i<j} A^{(-)}(i) A^{(-)}(j)\ldots A^{(+)}(n) + A^{(-)}(1)\ldots A^{(-)}(n) \end{aligned} \quad (142)$$

$$\begin{aligned} N(\varphi(1)\ldots\varphi(n)) = {} & \varphi^{(+)}(1)\ldots\varphi^{(+)}(n) + \sum_{i=1}^{n} \delta_p \varphi^{(-)}(i)\varphi^{(+)}(1)\ldots\varphi^{(+)}(n) \\ & + \sum_{i<j} \delta_p \varphi^{(-)}(i)\varphi^{(-)}(j)\ldots\varphi^{(+)}(n) + \varphi^{(-)}(1)\ldots\varphi^{(-)}(n) \end{aligned} \quad (143)$$

where $\delta_p = \pm 1$ according as $(ij\ldots)$ is an even or odd permutation of the sequence $(12\ldots n)$. We have obviously

$$N(\varphi(1)\ldots\varphi(n)) = \delta_p N(\varphi(i)\varphi(j)\ldots) \quad (144)$$

† The vacuum state is defined analogously to the lowest state of the harmonic oscillator (cf. p. 35).

Also for example

$$A(1)A(2) = N(A(1)A(2))+\langle 0'|A(1)A(2)|0'\rangle \tag{145}$$

$$\psi(1)\bar{\psi}(2) = N(\psi(1)\bar{\psi}(2))+\langle 0'|\psi(1)\bar{\psi}(2)|0'\rangle, \quad \text{etc.}$$

and

$$\tfrac{1}{2}[\bar{\psi}^{(\text{in})}(x), \gamma_\mu\psi^{(\text{in})}(x)] = N(\bar{\psi}^{(\text{in})}(x)\gamma_\mu\psi^{(\text{in})}(x)) \tag{146}$$

The S-matrix can therefore be written as mixed P–N products. Next define the T product of n operators by

$$T(A(1)\ldots A(n)) = P(A(1)\ldots A(n)) \tag{147}$$

and
$$T(\varphi(1)\ldots\varphi(n)) = \delta_p\varphi(i)\varphi(j)\ldots$$

where on the right-hand side the operators are ordered chronologically as in the P product. For example

$$T(A(1)A(2)) = \begin{cases} A(1)A(2) & \text{if } x_0^{(1)} > x_0^{(2)} \\ A(2)A(1) & \text{if } x_0^{(2)} > x_0^{(1)} \end{cases} \tag{148}$$

but

$$T(\varphi(1)\varphi(2)) = \begin{cases} \varphi(1)\varphi(2) & \text{if } x_0^{(1)} > x_0^{(2)} \\ -\varphi(2)\varphi(1) & \text{if } x_0^{(2)} > x_0^{(1)} \end{cases} \tag{149}$$

The relation between the T products and N products is given by

$$\begin{aligned} T(\varphi(1)\ldots\varphi(n)) &= N(\varphi(1)\ldots\varphi(n))+\textstyle\sum N(\varphi(1)\ldots\overbracket{\varphi(i)\ldots\varphi(j)}\ldots\varphi(n)) \\ &\quad +\textstyle\sum N(\varphi(1)\ldots\varphi(i_1)\ldots\varphi(i_2)\ldots\varphi(j_1)\ldots\varphi(j_2)\ldots\varphi(n))+\ldots \end{aligned} \tag{150}$$

(in the last sum $\varphi(i_1)$ is contracted with $\varphi(j_1)$ and $\varphi(i_2)$ with $\varphi(j_2)$)

where

$$N(\overbracket{\varphi(1)\varphi(2)}\varphi(3)\ldots\varphi(n)) = \langle 0'|T(\varphi(1)\varphi(2))|0'\rangle N(\varphi(3)\ldots\varphi(n)) \tag{151}$$

and

$$N(\varphi(1)\ldots\overbracket{\varphi(i)\ldots\varphi(j)}\ldots\varphi(n)) = \delta_p N(\overbracket{\varphi(i)\varphi(j)}\varphi(1)\ldots\varphi(n)) \tag{152}$$

The only non-vanishing vacuum expectations values of T products are

$$\left.\begin{aligned} \langle 0'|T(A_\mu^{(\text{in})}(x)A_\nu^{(\text{in})}(x'))|0'\rangle &= \tfrac{1}{2}\,\delta_{\mu\nu}D_F(x'-x) \\ \langle 0'|T(\bar{\psi}^{(\text{in})}(x)\psi^{(\text{in})}(x'))|0'\rangle &= \tfrac{1}{2}S_F(x'-x) \end{aligned}\right\} \tag{153}$$

We have, for example, also

$$N(\varphi(1)\ldots\overbracket{\varphi(i)\ldots\varphi(j)}\ldots\varphi(n)) = \delta_p N(\varphi(1)\ldots\overbracket{\varphi(j)\ldots\varphi(i)}\ldots(n))$$

and

$$P(N(\bar{\psi}(1)\psi(1))\ldots N(\bar{\psi}(n)\psi(n))) = T(N(\bar{\psi}(1)\psi(1))\ldots N(\bar{\psi}(n)\psi(n)))$$

The S-matrix contains therefore mixed T–N products. However, it is easy to see that there exists an expansion analogous to (150) for the mixed T–N products. For example

$$
\begin{aligned}
P\{N(\bar{\psi}(1)\psi(1))N(\bar{\psi}(2)\psi(2))\} \\
&= N(\bar{\psi}(1)\psi(1)\bar{\psi}(2)\psi(2))+\tfrac{1}{2}S_F(2,1)N(\psi(1)\bar{\psi}(2)) \\
&\quad -\tfrac{1}{2}S_F(1,2)N(\bar{\psi}(1)\psi(2))-\tfrac{1}{4}S_F(2,1)S_F(1,2) \qquad (154)
\end{aligned}
$$

Thus the integrands of all elements of the S-matrix can be expanded in a sum of normal products and terms containing normal products and the vacuum expectation values of T products. To compute a typical S-matrix element it is useful in practice to represent it graphically by means of so-called Feynman graphs. Thus, for the nth element of the S-matrix (after expansion in terms of normal products and the D_F or S_F functions) we represent each of the n space-time variables x by a point. In the expansion, for each term, each variable x appears three times as a result of the nature of the interaction. Each of the operators $A(i)$ appears as a factor in the normal product or together with $A(j)$ in $D_F(i, j)$. When two variables appear in the same function $D_F(i, j)$ we join the corresponding two points by a dotted line (internal). When $A(i)$ appears in a normal product we draw an open (external) dotted line from $x^{(i)}$. For the operators $\psi(i)$ or $\bar{\psi}(i)$ we do the same but this time with undotted lines. For example the Feynman graph for

$$
\begin{aligned}
\frac{(-e)^2}{2!}\iint dx\,dx'\langle q|P(N(\bar{\psi}^{(\text{in})}(x)\gamma_\mu\psi^{(\text{in})}(x))N(\bar{\psi}^{(\text{in})}(x')\gamma_\nu\psi^{(\text{in})}(x')))|q'\rangle \\
\times\langle 0'|P(A_\mu^{(\text{in})}(x)A_\nu^{(\text{in})}(x')|0'\rangle \\
= \frac{-e^2}{4}\iint dx\,dx'\langle q|\bar{\psi}^{(\text{in})}(x)|0'\rangle\gamma_\lambda S_F(x-x')\gamma_\lambda\langle 0'|\psi^{(\text{in})}(x')|q'\rangle D_F(x'-x) \\
\times D_F(x'-x) \qquad (155)
\end{aligned}
$$

is

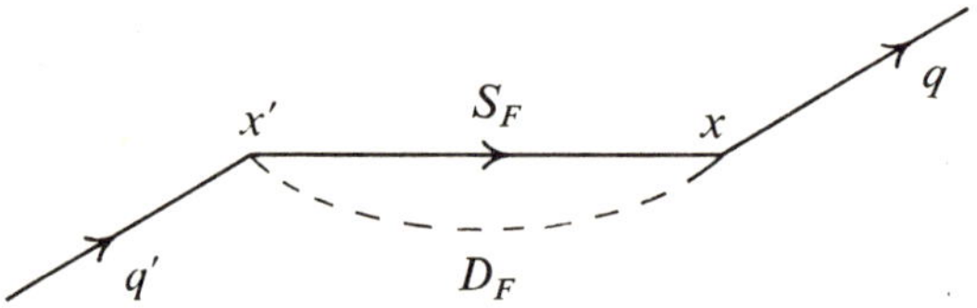

We have thus the following rule for the 1-1 correspondence between the graph and the corresponding matrix element (apart from multiplicative factors):

(a) Each internal dotted line corresponds to $(1/2)\,\delta_{\nu_i\nu_j}D_F(i, j)$.
(b) Each internal undotted line corresponds to $-(1/2)S_F(j, i)$.

(c) Each external line corresponds to a creation or a destruction operator according as the direction of the line.
(d) Each point $x^{(i)}$ corresponds to a factor γ_{ν_i}.

The primary usefulness of the graphs is that there also exists a 1-1 correspondence between the intuitive physical picture which a graph suggests and the actual physical process which one is interested in.

Re-normalization procedure. Although in principle one could compute the S-matrix elements of any order one gets into mathematical ambiguities in the form of infinities the moment higher order terms are taken into account. It has been possible to separate and classify the meaningful types of divergencies in quantum electrodynamics. These correspond to the following elementary Feynman graphs of second order:

(i) electron self-energy: 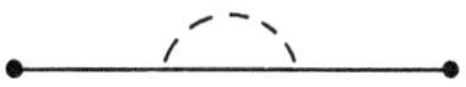

(ii) photon self-energy: 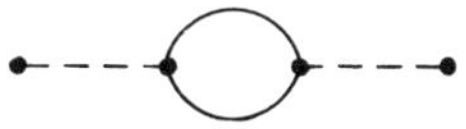

for the internal lines and corresponding energy graphs for the external lines and the

(iii) Vertex part: 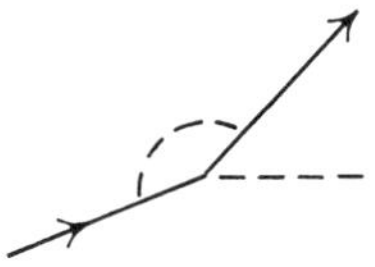

It is then possible to extract from these infinite quantities finite contributions and to get rid of the remaining infinite quantities in a unique and unambiguous manner by means of so-called mass and charge renormalizations.

Consider, for example, the effect of adding the electron self-energy graph to an internal electron line

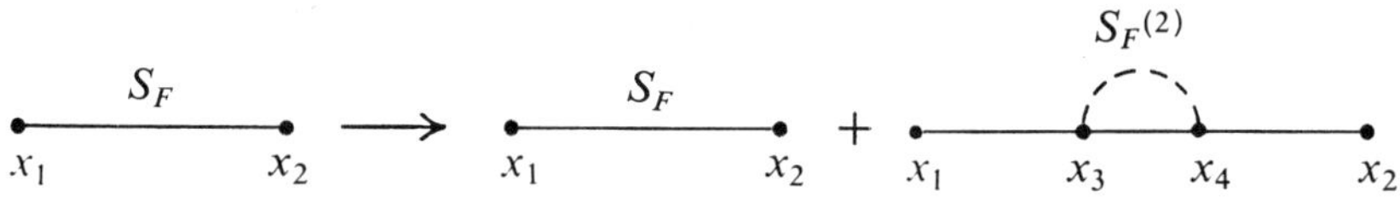

We write

$$
\begin{aligned}
S_F(x) &= \frac{1}{4\pi^3}\int dp\, e^{ipx} S_F(p) \\
D_F(x) &= \frac{1}{4\pi^3}\int dk\, e^{ikx} D_F(k)
\end{aligned}
\tag{156}
$$

The self-energy graph then corresponds to (apart from numerical factors).

$$e^2\int dx_3\int dx_4\, S_F(x_2-x_4)\gamma_\mu S_F(x_4-x_3)\gamma_\mu S_F(x_3-x_1)D_F(x_3-x_4) \tag{157}$$

In momentum space this corresponds to

$$S_F^{(2)}(p) = S_F(p)\sum(p)\,S_F(p) \tag{158}$$

where

$$\sum(p) = e^2\int dk\,\gamma_\mu S_F(p+k)\gamma_\mu D_F(k) \tag{159}$$

$\sum(p)$ happens to be divergent. However, it is possible to show that one can express $S^{(2)}(p)$ in the following manner

$$S_F^{(2)}(p) = S_F(p)AS_F(p) + \left(\frac{1}{2\pi}\right)BS_F(p) + \left(\frac{1}{2\pi}\right)S_C(p)S_F(p) \tag{160}$$

where A, B are infinite constants and $S_C(p)$ is a convergent quantity. This separation of an infinite quantity into a finite and infinite part—however dubious mathematically it may be—is a perfectly unambiguous procedure.† Thus the resultant contribution can be written as

$$S_F'(p) = S_F(p)+S_F^{(2)}(p) \tag{161}$$

If one now goes back to the field-equations and remembers that the mass m of the electron refers to the free-field mass, whereas the electron is always physically in interaction with the electromagnetic field, it seems appropriate to add a term δm to m. Thus we put

$$\left.\begin{array}{ll} & m = m+\delta m \\ \text{or} & m = m_0-\delta m \end{array}\right\} \text{ where } \left\{\begin{array}{l} \delta m = \text{self-mass} \\ m = \text{free mass} \\ m_0 = \text{observed mass} \end{array}\right. \tag{162}$$

Neither m nor δm are observable. The effect of δm is to introduce an extra term in the interacting Lagrangian, which in turn contributes to $S_F(p)$ a term

$$-2\pi i\,\delta m\, S_F(p)S_F(p) \tag{163}$$

From now on all the S_F functions contain the observed *re-normalized* mass m_0. Hence the corrected $S_F(p)$ becomes

$$S_F'(p) = S_F(p)+S_F(p)(A-2\pi i\,\delta m)S_F(p)+\left(\frac{1}{2\pi}\right)BS_F(p)+\left(\frac{1}{2\pi}\right)S_c(p)S_F(p) \tag{164}$$

† Dyson, F. J. (1949) *Phys. Rev.* **75**, 1736.

We can therefore get rid of A by equating it to the unobservable self-mass, that is, by putting

$$\delta m = A/2\pi i \tag{165}$$

The self-energy is therefore infinite as in classical electrodynamics. After mass re-normalization (i.e. removal of A) $S'_F(p)$ can be written, *up to second order*, as

$$S'_F(p) = Z_2 S_F(p) 1 + \left(\frac{1}{2\pi}\right) S_c(p) \tag{166}$$

where $Z_2 = 1 + B/2\pi$ is a re-normalization constant. The effect of adding a photon self-energy graph to an internal photon line is

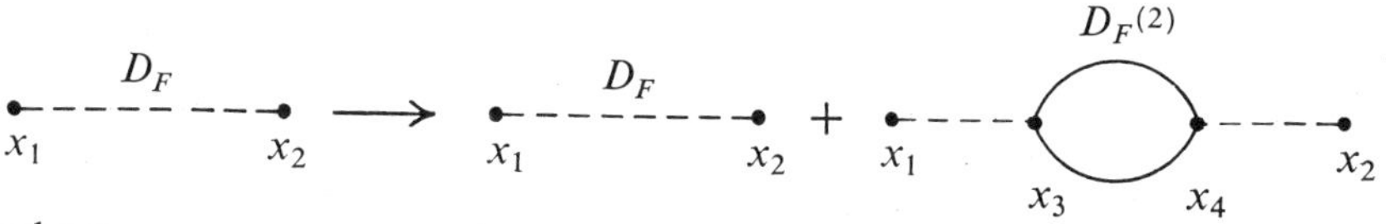

where

$$D_F^{(2)}(x_1 - x_2)_{\mu_1\mu_2}$$

$$= e^2 \iint dx_3\, dx_4\, D_F(x_1 - x_3) Tr\{\gamma_{\mu_1} S_F(x_3 - x_4)\gamma_{\mu_2} S_F(x_4 - x_3)\} D_F(x_4 - x_2) \tag{167}$$

or

$$D_F(p)_{\mu_1\mu_2} = D_F(p) \Pi_{\mu_1\mu_2}^{(p)} D_F(p)$$

where

$$\Pi_{\mu_1\mu_2}(p) = e^2 \int dk\, Tr\{\gamma_{\mu_1} S_F(p+k)\gamma_{\mu_2} S_F(k)\} \tag{168}$$

is again a divergent quantity. A similar procedure gives

$$\Pi_{\mu_1\mu_2}(p) = \delta_{\mu_1\mu_2} \Pi(p)$$
$$\Pi(p) = A' + Cp^2 + p^2 D_c(p) \tag{169}$$

Here A' corresponds to self-mass of the photon, which, however, vanishes because of zero rest mass of the photon. C is an infinite constant and $D_c(p)$ is convergent. Finally, one obtains for the corrected D_F-function up to second order

$$D'_F(p) = D_F(p) + D_F^{(2)}(p)$$
$$= Z_3 D_F(p)\left(1 + \left(\frac{1}{2\pi i}\right) D_c(p)\right) \tag{170}$$

where the re-normalization constant $Z_3 = 1 + \left(\frac{1}{2\pi i}\right) C$.

Consider now the addition of a vertex part in a graph shown below:

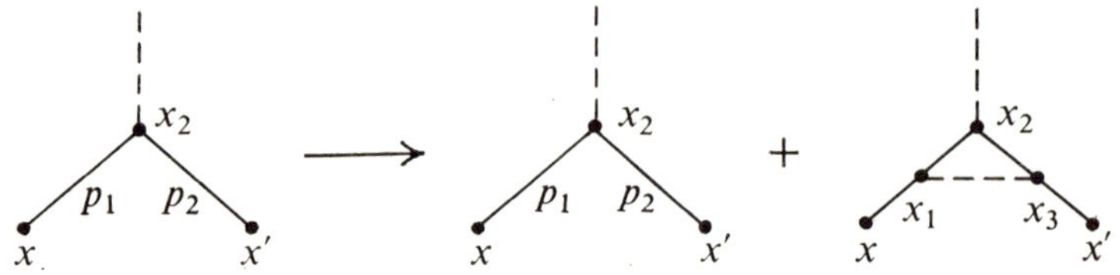

The initial graph corresponds to $S_F(p_1)\gamma_\mu S_F(p_2)$. The final graph corresponds to

$$S_F(p_1)\gamma_\mu S_F(p_2) + S_F(p_1)L_\mu(p_1, p_2)S_F(p_2)$$

where
$$L_\mu(p_1, p_2) = L\gamma_\mu + \Lambda_{\mu c}(p_1', p_2) \tag{171}$$

Here L is an infinite constant whereas $\Lambda_{\mu c}$ is finite. The corrected graph corresponds to

$$Z_1^{-1}S_F(p_1)(\gamma_\mu + \Lambda_{\mu c}(p_1, p_2))S_F(p_2) \tag{172}$$

where $Z_1 = 1 - L$. The vertex modification, however, does not introduce any new re-normalization constant because Z_1 *happens to be equal to* Z_2.

Re-normalization of external lines can be carried out in a similar fashion and one obtains

$$\psi'(p) = Z_2^{1/2}\psi(p) \tag{173}$$

corresponding to the graph

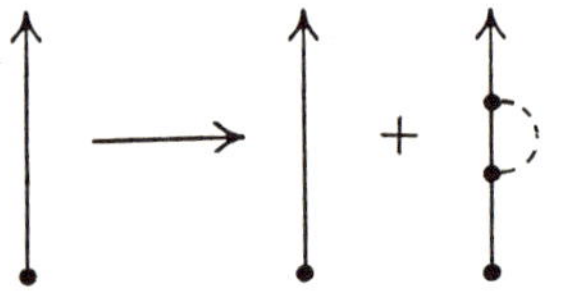

and $A_\mu'(p) = Z_3^{1/2}A_\mu(p)$ corresponding to

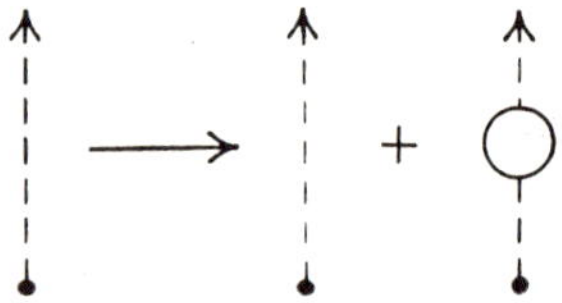

If one now considers an arbitrary graph containing n vertices, F_e internal electron lines, F_p internal photon lines, E_e external electron lines and E_p external photon lines, the matrix element contains the following infinite factor

$$Z_2^{E_e/2}Z_3^{E_p/2}Z_2^{F_e}Z_3^{F_p}Z_1^{-n} \tag{174}$$

It also contains the factor e^n, e being the charge. Since, however, $2F_e+E_e = 2n$, $2F_p+E_p = n$ and $Z_1 = Z_2$ we get finally a factor $e^n Z_3^{n/2}$. The remaining factors are finite. One now defines the re-normalized charge e_0 by $e_0 = eZ_3^{1/2}$ as the observable charge, e being the unobservable bare charge. It can be shown that this procedure works consistently in any order. The re-normalization procedure thus enables one to remove all the infinities by only charge and mass re-normalization. It, however, does not solve the basic problem of infinite self-energies; it merely succeeds in by-passing the problem. At one time it was thought that the infinities are a consequence of the perturbation method used in solving the field equations. According to Källen,† however, the difficulties are inherent in the theory quite independent of the perturbation method.

The theory also fails to provide a value for the masses of charged particles and the charge e_0 itself. In this connection the re-normalization procedure might have a deeper significance, in that, in a re-normalization theory with finite re-normalization constants Z_1, Z_2, so forth, one would be able to compute e_0 and δm (and therefore m_0 if all mass were self-mass) from only one or more fundamental parameters in the theory. Such a theory would probably require a radical change in the basic concepts of quantum field theory.

† Källen, G. (1952) *Helv. Phys. Acta.* **25**, 417; (1953) *Kgl. Danske Videnskab. Selskab.* **27**, 12.

Part 3

Unified Non-dualistic Theories

A. Classical Theories
B. Quantum Theory

A. Classical Theories

1. Theory of Einstein and Schrödinger

The problem of unification. The general theory of relativity differs from the special theory in one rather fundamental respect; it has introduced in physics a new principle: the *principle of geometrization of physics*. The general theory, in fact, succeeds in geometrizing the phenomenon of gravitation by abandoning the flat space-time of special theory and identifying the metric tensor of a Riemannian space-time with the gravitational potential.

If we want to treat also the electromagnetic field along with the gravitational field we can do so provisionally by adding the electromagnetic energy-momentum tensor $E_{\mu\lambda}$ to the right-hand side of Part 1, equation (74), and generalize Maxwell's equations by replacing ordinary derivatives by covariant derivatives:

$$\begin{aligned} R_{\mu\lambda}-\tfrac{1}{2}g_{\mu\lambda}R &= -\kappa(T_{\mu\lambda}+E_{\mu\lambda}) \\ f^{\mu\lambda}{}_{;\mu} &= s^{\lambda} \\ f_{\mu\lambda;\nu}+f_{\lambda\nu;\mu}+f_{\nu\mu;\lambda} &= 0 \end{aligned} \tag{1}$$

These are the Einstein-Maxwell's equations. The electromagnetic field, however, unlike the gravitational field has to be introduced from the outside and does not have any geometrical significance.

Ever since Einstein's theory of gravitation attempts were made to construct a so-called unified theory, in which both electromagnetism and gravitation could be incorporated into a single geometrical structure of a space-time manifold. Since all the ten components of the symmetric metric tensor of a Riemannian space-time are needed to describe the gravitational field, it is clear that unification in the above sense can only be achieved in the framework of either a generalization of the usual Riemannian space-time or even non-Riemannian space-times.

The first attempt in this direction was made by Weyl† in 1918. Recall that the geometrical structure of a manifold is determined by (i) an affine connection characterized by its components $\Gamma^{\mu}_{\alpha\beta}$ which are defined by the change due to infinitesimal parallel transfer of a vector ξ^{μ} from $P(x^{\mu})$ to $P'(x^{\mu}+dx^{\mu})$:

$$\delta\xi^{\mu} = -\Gamma^{\mu}_{\alpha\beta}\xi^{\alpha}\,dx^{\beta} \tag{2}$$

and (ii) a metrical connection characterized by the metric fundamental

† Weyl, H. (1918) *Sitz. Ber. d. Preuss. Akad. Wiss.* 465.

tensor $g_{\mu\lambda}$ which is defined by the measure of length ξ of a vector ξ^μ:

$$\xi = g_{\mu\lambda}\xi^\mu\xi^\lambda \tag{3}$$

Riemannian geometry is characterized by the symmetry of $\Gamma^\mu_{\alpha\beta}$ that is, $\Gamma^\mu_{\alpha\beta} = \Gamma^\mu_{\beta\alpha}$ and by the condition that the length of a vector should not change by a parallel transfer:

$$\delta\xi = \delta(g_{\mu\lambda}\xi^\mu\xi^\lambda) = 0 \tag{4}$$

By this condition the affine connection of a Riemannian manifold is uniquely determined by the metric tensor:

$$\Gamma^\mu_{\alpha\beta} = \left\{ {\mu \atop \alpha\beta} \right\} \tag{5}$$

where the right-hand side are Christoffel symbols of the second kind. Note the difference between (2) and (4). Weyl generalized the notion of the Riemannian metrical connection in the following manner.

With each vector ξ^μ there is an associated measure of length $\xi = g_{\mu\lambda}\xi^\mu\xi^\lambda$. Two vectors ξ^μ and η^μ have therefore the same length if and only if $\xi = \eta$. The quadratic form is completely determined only when one specifies a non-zero scale (or gauge) factor of proportionality. At every point of the manifold one has therefore the possibility of a change of scale (that is, a recalibration) or a *gauge transformation.* In a Weyl manifold it is not sufficient for the metrical connection to have a measure-determination at every point; every point must also be metrically related to the surrounding neighbourhood. The concept of metrical relationship is analogous to that of affine relationship. The metrical relationship is determined by specifying the change in the measure of length of a vector due to an infinitesimal parallel transfer. There exists a geodesic gauge in which there is no change in the measure of length of ξ^μ transferred parallelly from $P(x^\mu)$ to $P'(x^\mu + dx^\mu)$. In an arbitrary gauge, however, ξ is assumed to change (in contrast to (4)) according as

$$\delta\xi = -\xi\phi_\mu(x)\, dx^\mu \tag{6}$$

where $\phi_\mu(x)$ is a vector function characterizing the manifold. The metrical connection of a Weyl manifold is therefore characterized by *two* independent quantities $g_{\mu\lambda}$ and ϕ_μ, relative to a reference system (= co-ordinate system + gauge). If one makes a gauge transformation, that is, recalibrate all lengths: $\xi \to \bar{\xi} = \lambda\xi$, where $\lambda \equiv \lambda(x)$ in general, ϕ_μ and $g_{\mu\lambda}$ transform in the following manner:

$$g_{\mu\lambda} \to \bar{g}_{\mu\lambda} = \lambda g_{\mu\lambda}; \qquad \phi_\mu \to \bar{\phi}_\mu = \phi_\mu - \frac{1}{\lambda}\frac{\partial\lambda}{\partial x^\mu}$$

Analogous to the *vector curvature* of affine connection one can define a *length curvature* in the metrical connection by considering the change in the measure of length of an arbitrary vector ξ^μ around a closed infinitesimal parallelogram ABCD of sides dx_1^μ and dx_2^μ (cf. Part 1, equation (50)). If we displace parallelly from A to D via B first and then via C, the change in the measure of length is

$$\Delta\xi = -\xi f_{\mu\lambda}\, dx_1^\mu\, dx_2^\lambda$$

where

$$f_{\mu\lambda} = \phi_{\mu,\lambda} - \phi_{\lambda,\mu} \tag{7}$$

is the so-called length curvature of the Weyl manifold. A *necessary* and *sufficient* condition for a length to be transported parallelly from one point to another independent of the path chosen is

$$f_{\mu\lambda} = 0$$

A Riemannian manifold is thus a special case of the Weyl manifold with vanishing length curvature.

The components of affine connection $\Gamma^\mu_{\alpha\beta}$ of a Weyl manifold are now determined through (2) and (6) by $g_{\mu\lambda}$ and ϕ_μ:

$$\Gamma^\mu_{\alpha\beta} = \left\{ {\mu \atop \alpha\beta} \right\} + \tfrac{1}{2}(\delta^\mu_\alpha \phi_\beta + \delta^\mu_\beta \phi_\alpha - g_{\beta\alpha}\phi^\mu) \tag{8}$$

where $\phi^\mu = g^{\mu\lambda}\phi_\lambda$.

As usual the vector curvature is given by

$$F_{\kappa\alpha\beta}^{\ \ \mu} = \Gamma_{\kappa\beta,\alpha}^{\ \mu} - \Gamma_{\kappa\alpha,\beta}^{\ \mu} + \Gamma^\mu_{\rho\alpha}\Gamma^\rho_{\kappa\beta} - \Gamma^\mu_{\rho\beta}\Gamma^\rho_{\kappa\alpha} \tag{9}$$

Substituting (8) in (9)

$$F_{\kappa\alpha\beta}^{\ \ \mu} = R_{\kappa\alpha\beta}^{\ \ \mu} + S_{\kappa\beta;\alpha}^{\ \mu} - S_{\kappa\alpha;\beta}^{\ \mu} + S^\mu_{\alpha\rho}S^\rho_{\kappa\beta} - S^\mu_{\beta\rho}S^\rho_{\kappa\alpha}$$

where $R_{\kappa\alpha\beta}^{\ \ \mu}$ is the Riemannian curvature formed with the Christoffel symbols $\left\{ {\mu \atop \alpha\beta} \right\}$ and

$$S^\mu_{\alpha\beta} = \Gamma^\mu_{\alpha\beta} - \left\{ {\mu \atop \alpha\beta} \right\}$$

$$S^\mu_{\kappa\beta;\alpha} \equiv \text{covariant derivative of } S^\mu_{\kappa\beta} \text{ relative to } \left\{ {\mu \atop \alpha\beta} \right\}$$

The contracted curvature tensor is given by

$$F_{\kappa\alpha} \equiv F_{\kappa\alpha\mu}^{\ \ \mu} = R_{\kappa\alpha} + \tfrac{3}{2}\phi_{\kappa;\alpha} - \tfrac{1}{2}\phi_{\alpha;\kappa} + \tfrac{1}{2}g_{\alpha\kappa}\phi^\mu_{;\mu} + \tfrac{1}{2}g_{\alpha\kappa}\phi^\lambda\phi_\lambda + \tfrac{1}{2}\phi_\kappa\phi_\alpha \tag{10}$$

and the curvature scalar is

$$F \equiv g^{\kappa\alpha}F_{\kappa\alpha} = R + \tfrac{3}{2}\phi_\lambda\phi^\lambda + 3\phi^\mu_{;\mu} \tag{11}$$

The vector function ϕ_μ can now be identified with the vector potential of electromagnetic field. The first set of Maxwell's equations

$$f_{\mu\lambda,\nu} + f_{\lambda\nu,\mu} + f_{\nu\mu,\lambda} = 0 \tag{12}$$

then follows immediately from (7). $g_{\mu\lambda}$, as usual, signifies the gravitational potential. To derive the remaining field equations one proceeds most naturally from a variational principle

$$\delta \int \mathscr{W} \, dv = 0 \tag{13}$$

where $\mathscr{W}$ is a scalar density under co-ordinate transformations but *invariant* under gauge transformations. $g_{\mu\lambda}$ and ϕ_μ being independent variables whose variations $\delta g_{\mu\lambda}$ and $\delta\phi_\mu$ vanish at the boundary, we can write

$$\delta \int \mathscr{W} \, dv \equiv \int (\mathscr{W}^{\mu\lambda} \, \delta g_{\mu\lambda} + \omega^\lambda \, \delta\phi_\lambda) \, dv = 0 \tag{14}$$

from which follow the field equations:

$$\mathscr{W}^{\mu\lambda} = 0 \tag{15}$$

$$\omega^\lambda = 0 \tag{16}$$

The invariance of the action integral under co-ordinate as well as gauge transformations furnishes us five identities for $\mathscr{W}^{\mu\lambda}$ and ω^λ. Consider for example an infinitesimal gauge transformation

$$\xi \rightarrow \lambda\xi; \; \lambda = 1 + \delta\lambda \tag{17}$$

The changes in $g_{\mu\lambda}$ and ϕ_λ are

$$\left.\begin{aligned} \delta g_{\mu\lambda} &= g_{\mu\lambda} \, \delta\lambda \\ \delta\phi_\mu &= -\frac{\partial(\delta\lambda)}{\partial x^\mu} \end{aligned}\right\} \tag{18}$$

Substituting (18) in (14) and after a partial integration we get

$$\int (\omega^\mu{}_{,\mu} - \mathscr{W}^\mu_\mu) \, \delta\lambda \, dv = 0; \qquad \mathscr{W}^\mu_\mu \equiv \mathscr{W}^{\mu\lambda} g_{\mu\lambda}$$

Since $\delta\lambda$ is arbitrary we get the identity

$$\omega^\mu{}_{,\mu} - \mathscr{W}^\mu_\mu = 0 \tag{19}$$

The four identities which result from invariance under co-ordinate transformations are

$$\mathscr{W}^\mu_{\lambda,\mu} - \mathscr{W}^\mu_\nu \Gamma^\nu_{\mu\lambda} - \tfrac{1}{2}\omega^\mu f_{\mu\lambda} = 0 \tag{20}$$

There are several possibilities for $\mathscr{W}$

$$F_{\mu\lambda\nu\kappa} F^{\mu\lambda\nu\kappa} \sqrt{-g}; \qquad F_{\mu\lambda} F^{\mu\lambda} \sqrt{-g}; \qquad F^2 \sqrt{-g}; \qquad f_{\mu\lambda} f^{\mu\lambda} \sqrt{-g}$$

Weyl considered the following simple combination

$$\mathscr{W} = (F^2 - \alpha f_{\mu\lambda} f^{\mu\lambda})\sqrt{-g} \tag{21}$$

where α = constant.

Pauli† has considered a more general combination

$$\mathscr{W} = (a_1 F_{\mu\lambda\nu\kappa} F^{\mu\lambda\nu\kappa} + a_2 F_{\mu\lambda} F^{\mu\lambda} + a_3 F^2 + a_4 f_{\mu\lambda} f^{\mu\lambda})\sqrt{-g}$$

with constants a_1, a_2, etc.

The field equations (15) and (16), which follow from Weyl's action function (21), contain in general derivatives of $g_{\mu\lambda}$ higher than second order. Weyl therefore introduced the gauge condition

$$F = \beta = \text{constant} \tag{22}$$

Consequently $\delta(F^2\sqrt{-g}) \equiv 2\beta\delta(F\sqrt{-g}) - \beta^2\delta(\sqrt{-g})$ and the variational principle leads to

$$\delta(R - \lambda - \varepsilon f_{\mu\lambda} f^{\mu\lambda} + \tfrac{3}{4}\beta\phi_\mu\phi^\mu)\sqrt{-g} = 0$$

where $\lambda = \beta/2$, $\varepsilon = \alpha/2\beta$. The field equations then become

$$R_{\mu\lambda} - \tfrac{1}{2}g_{\mu\lambda}R - \lambda g_{\mu\lambda} + \tfrac{3}{4}\beta(\phi_\mu\phi_\lambda - \tfrac{1}{2}g_{\mu\lambda}\phi_\nu\phi^\nu) + 2\varepsilon E_{\mu\lambda} = 0 \tag{23}$$

$$4\varepsilon \frac{\partial(f^{\mu\lambda}\sqrt{-g})}{\sqrt{(-g)}\,\partial x^\lambda} + \tfrac{3}{2}\beta\phi^\mu = 0 \tag{24}$$

where $E_{\mu\lambda}$ is the electromagnetic energy-momentum tensor. Equations (23) are the modified field-equations of gravitation in presence of an electromagnetic field. Note the cosmological term $\lambda g_{\mu\lambda}$ arising as a consequence of Weyl's gauge condition. A comparison of (24) with the Maxwell's second set of equations

$$f^{\mu\lambda}_{;\,\lambda} = s^\mu \equiv \text{current vector}$$

gives the remarkable result

$$s^\mu = -\tfrac{3}{4}\alpha\phi^\mu \tag{25}$$

that is, the current vector is proportional to the potential vector. As a consequence of the field-equations (15) and (16), the identities (19) and (20) become

$$\omega^\mu_{;\,\mu} = 0 \tag{26}$$

$$\mathscr{W}^\mu_{\lambda,\,\mu} - \Gamma^\mu_{\lambda\nu}\omega^\nu_\mu = 0 \tag{27}$$

(26), (25) and (24) imply the conservation of charge-current

$$s^\mu_{,\,\mu} = 0 \tag{28}$$

† Pauli, W. (1919) *Phys. Z.* **20**, 457.

It is thus possible to give a geometrical significance to the electromagnetic field and derive the Maxwell's equations in a straightforward manner. Weyl's theory is, however, subject to the following criticism according to Einstein. The hypothesis of non-integrability of length transfer implies that the frequency of spectral lines emitted by atoms would not remain constant but would depend on their past history. From the mathematical point of view the choice of Weyl's action function remains somewhat arbitrary and it is necessary to introduce a gauge condition to avoid field equations involving higher order derivatives.

In 1951 Lyra† suggested a modification of Riemannian geometry which may also be considered as a modification of Weyl's geometry. In Lyra's geometry Weyl's concept of gauge, which was essentially a *metrical* concept, is modified by the introduction of a gauge function in the *structureless* manifold.

According to Lyra, the displacement vector **PP′** between two neighbouring points $P(x^\mu)$ and $P'(x^\mu + dx^\mu)$ has the components $\xi^\mu = x^0\, dx^\mu$, where $x^0(x^\mu)$ is a gauge function. The co-ordinate system (x^μ) together with the gauge x^0 form a *reference system* $(x^0; x^\mu)$. The transformation formula for a tensor under a general transformation of reference systems

$$x^\mu \to x^{\mu'} = x^{\mu'}(x^\lambda); \qquad x^0 \to x^{0'} = x^{0'}(x^0, x^\mu) \tag{29}$$

with

$$A^{\mu'}_{\mu} \equiv \partial x^{\mu'}/\partial x^\mu; \qquad |A^{\mu'}_{\mu}| \neq 0, \quad \frac{\partial x^{0'}}{\partial x^0} \neq 0$$

is then given by

$$\xi^{\rho'_1 \cdots\cdots \rho'_s}_{\sigma'_1 \cdots\cdots \sigma'_n} = \lambda^{s-r} A^{\rho'_1}_{\rho_1} \cdots\cdots A^{\rho'_s}_{\rho_s} A^{\sigma_1}_{\sigma'_1} \cdots\cdots A^{\sigma_r}_{\sigma'_r} \xi^{\rho_1 \cdots \rho_s}_{\sigma_1 \cdots \sigma_r} \tag{30}$$

The factor λ^{s-r}, where $\lambda \equiv x^{0'}/x^0$ arises as a consequence of the introduction of the gauge function.

In an affinely connected manifold the components of the affine connection $\Gamma^\mu_{\alpha\beta}$ can be considered to arise as a consequence of general co-ordinate transformations in the following manner. Let us suppose that, in a co-ordinate system (x^μ), a vector ξ^μ is constant, that is, $\xi^\mu{}_{,\lambda} = 0$. Then, in another arbitrary co-ordinate system $(x^{\mu'})$, we have

$$\xi^{\mu'}_{,\lambda'} + \Gamma^{\mu'}_{\nu'\lambda'}\xi^{\nu'} = 0 \tag{31}$$

where

$$\Gamma^{\mu'}_{\nu'\lambda'} \equiv -A^{\mu}_{\nu'} A^{\mu'}_{\mu,\lambda'}; \qquad A^{\mu'}_{\mu,\lambda'} \equiv \frac{\partial}{\partial x^{\lambda'}}(A^{\mu'}_{\mu})$$

† Lyra, G. (1951) *Math. Z.* **54**, 52.

Another way of expressing the fact that ξ^μ is constant would be to say that equation (31) is valid in all co-ordinate systems, but that $\Gamma^\mu_{\nu\lambda} = 0$ in the original co-ordinate system (x^μ). The transition from an integrable affinely connected manifold to a non-integrable one (thereby from ordinary derivative to covariant derivative) is made by assuring that $\Gamma^\mu_{\nu\lambda} \neq 0$ in general.

Now a vector ξ^μ in Lyra's geometry transforms as

$$\xi^{\mu'} = \lambda A^{\mu'}_{\mu} \xi^\mu \tag{32}$$

If $\xi^\mu_{,\lambda} = 0$ in the reference system $(x^0; x^\mu)$, then, in the reference system $(x^{0'}; x^{\mu'})$ we have

$$\frac{1}{x^{0'}} \xi^{\mu'}_{,\lambda'} + \Gamma^{\mu'}_{\nu'\lambda'}\xi^{\nu'} - \tfrac{1}{2}\phi_{\lambda'}\xi^{\mu'} = 0 \tag{33}$$

where

$$\Gamma^{\mu'}_{\nu'\lambda'} = \left(\frac{-1}{x^{0'}}\right) A^\mu_{\nu'} A^{\mu'}_{\mu,\lambda'}; \qquad \phi_{\lambda'} = \left(\frac{1}{x^{0'}}\right)\frac{\partial \log \lambda^2}{\partial x^{\lambda'}}$$

The parallel transfer of a vector ξ^μ in Lyra's geometry is therefore given by

$$\delta\xi^\mu = -\tilde{\Gamma}^\mu_{\alpha\beta}\xi^\alpha x^0\, dx^\beta \tag{34}$$

where

$$\tilde{\Gamma}^\mu_{\alpha\beta} = \Gamma^\mu_{\alpha\beta} - \tfrac{1}{2}\delta^\mu_\alpha \phi_\beta$$

$$\tilde{\Gamma}^\mu_{\alpha\beta} \neq \tilde{\Gamma}^\mu_{\beta\alpha} \quad \text{but} \quad \Gamma^\mu_{\alpha\beta} = \Gamma^\mu_{\beta\alpha}$$

The components of the generalized affine connection are thus characterized not only by $\Gamma^\mu_{\alpha\beta}$ but also by ϕ_β which appear as a natural consequence of the formal introduction of the gauge function in the structureless space.

The covariant derivative of a mixed tensor is given by the usual formula with $\partial/\partial x^\mu$ replaced by $(1/x^0)(\partial/\partial x^\mu)$ and $\Gamma^\mu_{\alpha\beta}$ by $\tilde{\Gamma}^\mu_{\alpha\beta}$.

The metric or length of the displacement vector $\xi^\mu = x^0\, dx^\mu$ between two points $P(x^\mu)$ and $P'(x^\mu + dx^\mu)$ is defined by the absolute invariant (that is, invariant under gauge and co-ordinate transformations)

$$ds^2 = g_{\mu\lambda} x^0\, dx^\mu x^0\, dx^\lambda \tag{35}$$

where $g_{\mu\lambda}$ is a symmetric tensor of second rank. The parallel transfer of length in Lyra's geometry is integrable (as in Riemannian geometry) in contrast to Weyl's geometry, that is

$$\delta(g_{\mu\lambda}\xi^\mu\xi^\lambda) = 0 \tag{36}$$

From (34) and (36) it follows that

$$\Gamma^\mu_{\alpha\beta} = (1/x^0)\left\{{\mu \atop \alpha\beta}\right\} + \tfrac{1}{2}(\delta^\mu_\alpha \phi_\beta + \delta^\mu_\beta \phi_\alpha - g_{\alpha\beta}\phi^\mu) \tag{37}$$

It should be noted that apart from the factor $1/x^0$, (37) is identical with the components of affine connection in Weyl's geometry, not, however, $\tilde{\Gamma}^{\mu}_{\alpha\beta}$.

The parallel transfer (34) and therefore the system of differential equations

$$(1/x^0)\xi^{\mu}_{,\lambda}+\tilde{\Gamma}^{\mu}_{\alpha\lambda}\xi^{\alpha} = 0$$

is integrable only when the components of Lyra's curvature tensor

$$K^{\mu}_{\kappa\alpha\beta} = \left[\frac{1}{(x^0)^2}\right]\left[\frac{\partial(x^0\tilde{\Gamma}^{\mu}_{\kappa\beta})}{\partial x^{\alpha}}-\frac{\partial(x^0\tilde{\Gamma}^{\mu}_{\kappa\alpha})}{\partial x^{\beta}}+x^0\tilde{\Gamma}^{\mu}_{\rho\alpha}x^0\tilde{\Gamma}^{\rho}_{\kappa\beta}-x^0\Gamma^{\mu}_{\rho\beta}x^0\tilde{\Gamma}^{\rho}_{\kappa\alpha}\right] \quad (38)$$

identically vanish. The contracted curvature tensor $K_{\kappa\alpha} \equiv K^{\mu}_{\kappa\alpha\mu}$ can be split into a symmetric and antisymmetric part ($K_{\underline{\kappa\alpha}} = K_{\underline{\alpha\kappa}}$; $K_{\check{\kappa\alpha}} = -K_{\check{\alpha\kappa}}$)

$$K_{\underline{\kappa\alpha}} = \frac{R_{\kappa\alpha}}{(x^0)^2}+\frac{1}{2x^0}(\phi_{\kappa;\alpha}+\phi_{\alpha;\kappa})+\frac{1}{2x^0}g_{\alpha\kappa}\phi^{\mu}_{;\mu}-\tfrac{1}{2}\phi_{\alpha}\phi_{\kappa}+\tfrac{1}{2}g_{\alpha\kappa}\phi^{\lambda}\phi_{\lambda}+\tfrac{1}{2}g_{\alpha\kappa}\mathring{\phi}_{\mu}\phi^{\mu} \quad (39)$$

$$K_{\check{\kappa\alpha}} = \frac{f_{\kappa\alpha}}{2x^0}-\tfrac{3}{4}(\mathring{\phi}_{\kappa}\phi_{\alpha}-\mathring{\phi}_{\alpha}\phi_{\kappa})$$

where

$$\mathring{\phi}_{\alpha} = \frac{1}{x^0}\frac{\partial \log x^{0^2}}{\partial x^{\alpha}}$$

and $R_{\kappa\alpha}$ is the contracted Riemannian curvature tensor formed from the Christoffel symbols. The curvature scalar in Lyra's geometry is then

$$K \equiv g^{\kappa\alpha}K_{\kappa\alpha} = \frac{R}{(x^0)^2}+\frac{3}{x^0}\phi^{\mu}_{;\mu}+\tfrac{3}{2}\phi_{\mu}\phi^{\mu}+2\mathring{\phi}_{\mu}\phi^{\mu} \quad (40)$$

which reduces to the Weyl curvature scalar F (see equation (11)) if one adopts the *normal* gauge $x^0 \equiv 1$.

It is therefore possible to construct a unified field theory in the framework of Lyra's geometry, with almost identical results as in Weyl's theory.† A variational principle in Lyra's geometry would have the form

$$\delta\int \mathscr{W}\,dv = \delta\int W\sqrt{(-g)}\,x^0\,dx^1\ldots x^0\,dx^4 = 0 \quad (41)$$

where W is a scalar function of $g_{\mu\lambda}$, ϕ_{μ} and x^0. A simple choice for W is to consider (e.g. in the normal gauge $x^0 = 1$)

$$\delta\int (K-\alpha f_{\mu\lambda}f^{\mu\lambda})\sqrt{(-g)}\,dv = 0 \quad (42)$$

† Sen, D. K. (1958) Thèse, Paris.

from which follow the following field equations:

$$R_{\mu\lambda}-\tfrac{1}{2}g_{\mu\lambda}R+3\phi_{\mu}\phi_{\lambda}-\tfrac{3}{4}g_{\mu\lambda}\phi_{\nu}\phi^{\nu}+8\pi\alpha E_{\mu\lambda} = 0 \tag{43}$$

$$\frac{\partial(f^{\mu\lambda}\sqrt{-g})}{\sqrt{(-g)}\,\partial x^{\lambda}}+\tfrac{3}{4}\alpha\phi^{\mu} = 0 \tag{44}$$

Apart from a cosmological term and constant factors (43) and (44) are identical with Weyl's field equations. However, we do not have here the difficulty of non-integrability of length transfer. Consequently, the most serious objection to Weyl's theory is removed. Moreover, it is not necessary to impose *ad hoc* gauge conditions (see equation (22)) to avoid higher order field equations.

Besides the Weyl type of unified theories, there have been two major different approaches to the problem of unification.

One way is to increase the number of dimensions of the manifold from four to five without essentially changing the Riemannian character of the geometry.†

Consider the motion of a charged particle of mass m and charge e in the combined gravitational and electromagnetic field characterized by the potentials $g_{\mu\lambda}$ and ϕ_{α}. In a four-dimensional Reimannian space-time V_4 the trajectories are no longer geodesics but are given by the generally covariant Lorentz equations of motion:

$$\frac{d^2x^{\beta}}{ds^2}+\left\{{\beta \atop \lambda\mu}\right\}\frac{dx^{\lambda}}{ds}\frac{dx^{\mu}}{ds} = \frac{e}{m}f^{\beta}_{\ \alpha}\frac{dx^{\alpha}}{ds} \tag{45}$$

$$\mu, \lambda = 1, 2, 3, 4$$

The right-hand side represents the covariant Lorentz force. Equation (45) can be derived from the variational principle:

$$\delta\int L(x, \dot{x})\,du = 0$$
$$L = (g_{\mu\lambda}\dot{x}^{\mu}\dot{x}^{\lambda})^{1/2}+k\phi_{\mu}\dot{x}^{\mu} \tag{46}$$

where $k = e/m$, $\dot{x}^{\mu} \equiv dx^{\mu}/du$, and u is an arbitrary parameter.

By suitably modifying the geometric structure of the manifold one could consider (45) as geodesics of the modified geometry. For example, (45) can be considered as geodesics of a four-dimensional Finsler space. But then, with each type of particle, that is, each value of k one must associate a different Finsler space. On the other hand, if we do not wish to destroy the Riemannian character of the manifold, (45) can be considered as geodesics of a five-dimensional Riemannian manifold V_5 with a metric:

$$d\sigma^2 = \gamma_{ik}\,dx^i\,dx^k; \qquad i, k = 0, 1, 2, 3, 4\ldots \tag{47}$$

† Kaluza, T. (1921) *Sitz. Ber. d. Preuss. Akad. Wiss.* 966. Klein, O. (1926) *Z. Phys.* **37**, 895.

The symmetric metric tensor γ_{ik} now has 15 components. In order to identify them with the 14 physical components $g_{\mu\lambda}$ and ϕ_α one must make some special assumptions. First of all, the four co-ordinates x^μ ($\mu = 1, 2, 3, 4$) must always represent the physical space-time co-ordinates. Second, γ_{ik} must be *independent* of x^0. This implies that the permissible co-ordinate transformations are to be confined to the following group:

$$\begin{aligned} x^0 &\to x^{0'} = x^0 + f^0(x^\mu) \\ x^\mu &\to x^{\mu'} = f^\mu(x^\mu) \end{aligned} \tag{48}$$

with arbitrary $f^i(x^\mu)$.

Under the transformation (48) γ_{00} remains invariant and therefore we can take

$$\gamma_{00} = \alpha \text{ (constant)} \tag{49}$$

Moreover, the transformation leaves the following differential expressions invariant

$$d\theta \equiv dx^0 + \frac{\gamma_{0\mu}}{\gamma_{00}} dx^\mu \tag{50}$$

$$ds^2 \equiv \left(\gamma_{\mu\lambda} - \frac{\gamma_{0\mu}\gamma_{0\lambda}}{\gamma_{00}}\right) dx^\mu \, dx^\lambda \tag{51}$$

$d\theta$ and ds^2 are related to the metric $d\sigma^2$ as follows:

$$d\sigma^2 = \alpha \, d\theta^2 + ds^2 \tag{52}$$

In view of the invariance of $d\theta$ and γ_{00} under (48) it follows that the four components $\gamma_{0\mu}$ transform as a 4-vector if x^0 is *kept fixed* and only the x^μ are transformed. On the other hand, if x^0 is also transformed along with x^μ, according to (48), an additive gradient of a scalar function appears. In other words $\gamma_{0\mu}$ behave as the electromagnetic potentials. One, therefore, sets

$$\begin{aligned} \gamma_{0\mu} &= \alpha\beta\phi_\mu \\ d\theta &= dx^0 + \beta\phi_\mu \, dx^\mu \end{aligned} \tag{53}$$

where β is another constant, and

$$g_{\mu\lambda} = \gamma_{\mu\lambda} - \frac{\gamma_{0\mu}\gamma_{0\lambda}}{\gamma_{00}}$$

or

$$\gamma_{\mu\lambda} = g_{\mu\lambda} + \alpha\beta^2\phi_\mu\phi_\lambda \tag{54}$$

Equation (47) therefore becomes

$$d\sigma^2 = g_{\mu\lambda} \, dx^\mu \, dx^\lambda + \alpha(dx^0 + \beta\phi_\mu \, dx^\mu)^2$$

The geodesics can now be interpreted as trajectories of charged particles. From the independence of γ_{ik} of x^0 it follows that for geodesics we may take

$$\frac{dx^0}{ds}+\beta\phi_\mu\frac{dx^\mu}{d\sigma} = \text{const.}$$

$$g_{\mu\lambda}\frac{dx^\mu}{d\sigma}\frac{dx^\lambda}{d\sigma} = \text{const.}$$

for a suitable choice of the parameter σ. We have therefore

$$\frac{d^2x^\mu}{d\sigma^2}+\left\{{\mu \atop \alpha\beta}\right\}\frac{dx^\alpha}{d\sigma}\frac{dx^\beta}{d\sigma} = \text{const}\, f^\mu_\lambda\frac{dx^\lambda}{d\sigma}$$

which can be compared with (45).

The field-equations in vacuum are to follow from the usual variational principle

$$\delta\int R\sqrt{(-\gamma)}\,dx^0\,dx^1\ldots dx^4 = 0 \tag{55}$$

where R is the curvature scalar in V_5 and $\gamma \equiv |\gamma_{ik}|$. Here $R \equiv R(\gamma_{ik};\gamma_{ik,l})$ and $\gamma_{ik,0} = 0$, $\gamma_{00} = \alpha$. The field-equations are:

$$R^{\mu\lambda}-\frac{1}{2}g^{\mu\lambda}R+\frac{\alpha\beta^2}{2}E^{\mu\lambda} = 0 \tag{56}$$

$$\frac{\partial(\sqrt{(-g)}\,f^{\mu\lambda})}{\partial x^\mu} = 0 \tag{57}$$

A comparison with the Maxwell-Einstein equations (1) shows that we must set

$$\frac{\alpha\beta^2}{2} = \kappa \tag{58}$$

It is thus possible to put the Einstein-Maxwell theory formally in a unified 5-dimensional geometrical framework without, however, modifying its essential substance. Veblen† has shown that the Einstein-Maxwell theory can also be considered as a "projective theory of relativity" where the underlying geometry is that of a 5-dimensional projective manifold. This becomes clear when one considers the underlying group structure of the Einstein-Maxwell theory.‡

† Veblen, O. (1933) "Projektive Relätivitätstheorie", Berlin. Ludwig, G. (1951) "Fortschritte der projektiven Relätivitätstheorie", Braunschweig.

‡ Jordan, P. (1945) "Göttinger Nachr.", 74.

Let $\mathscr{C}$ denote the group of co-ordinate transformations

$$x^{\mu} \rightarrow x^{\mu'} = x^{\mu'}(x^{\nu}); \qquad A^{\mu'}_{\nu} \equiv \frac{\partial x^{\mu'}}{\partial x^{\nu}}; \qquad A^{\nu}_{\mu'} \equiv \frac{\partial x^{\nu}}{\partial x^{\mu'}} \tag{59}$$

under which the electromagnetic potentials transform as follows:

$$\phi_{\mu}(x) \rightarrow \phi_{\mu'}(x') = A^{\nu}_{\mu'}\phi_{\nu}(x) \tag{60}$$

And let $\mathscr{G}$ be the group of gauge transformations (see Part 1, equation (13)):

$$\phi_{\mu}(x) \rightarrow \bar{\phi}_{\mu}(x) = \phi_{\mu}(x) + \varphi(x)_{,\mu} \tag{61}$$

where φ is a scalar function of x^{μ}. If we make a co-ordinate transformation (59) first and then a gauge transformation (61) we obtain:

$$\begin{aligned} \phi_{\mu}(x) \rightarrow \phi_{\mu'}(x') \rightarrow \bar{\phi}_{\mu'}(x') &= \phi_{\mu'}(x') + \varphi(x')_{,\mu'} \\ &= \phi_{\nu}(x)A^{\nu}_{\mu'} + \varphi(x')_{,\mu'} \end{aligned} \tag{62}$$

If we now go back to the original co-ordinate system, that is, make the inverse transformation to (59), we obtain

$$\begin{aligned} \bar{\phi}_{\mu'}(x') \rightarrow \bar{\bar{\phi}}_{\mu}(x) = A^{\lambda'}_{\mu}\bar{\phi}_{\lambda'}(x') &= A^{\lambda'}_{\mu}(A^{\nu}_{\lambda'}\phi_{\nu}(x) + \varphi(x')_{,\lambda'}) \\ &= \phi_{\mu}(x) + \varphi(x'(x))_{,\mu} \end{aligned}$$

which is again a gauge transformation.

Thus for $c \in \mathscr{C}$ we have $c^{-1}\mathscr{G}c = \mathscr{G}$. Now the Einstein-Maxwell equations are invariant under the combined group of gauge and co-ordinate transformations, which we denote by $\mathscr{K}$. Thus the gauge group $\mathscr{G}$ is an *invariant* subgroup of $\mathscr{K}$, not, however, $\mathscr{C}$. The group $\mathscr{K}$ is isomorphic to $\mathscr{H}$, the group of all transformations:

$$X^{i} \rightarrow X^{i'} = X^{i'}(X^{k}); \qquad i, k = 0, 1, 2, 3, 4$$

such that $X^{i'}$ *are homogeneous functions of degree one*, that is

$$X^{i'}_{,k}X^{k} = X^{i'}; \qquad X^{i'}_{,k} \equiv \frac{\partial X^{i'}}{\partial X^{k}} \tag{63}$$

This can be seen as follows. The general element of $\mathscr{H}$ is of the form:

$$X^{i} \rightarrow X^{i'} = X^{i}F^{(i)}(X^{1}/X^{0}, X^{2}/X^{0}, X^{3}/X^{0}, X^{4}/X^{0}) \tag{64}$$

where $F^{(i)}$ are arbitrary functions and there is no summation over i. The subgroup of all transformations where all the five functions $F^{(i)}$ are same, that is, $F^{(i)} = F$ (for *all* i) form an *abelian* invariant subgroup of $\mathscr{H}$, which can be mapped onto the gauge group $\mathscr{G}$ as follows:

$$F \equiv e^{\varphi} \tag{65}$$

On the other hand $\mathscr{C}$, the group of co-ordinate transformations (59) is mapped onto the following subgroup of transformations of $\mathscr{H}$:

$$X^0 \to X^{0'} = X^0$$

$$X^\mu \to X^{\mu'} = X^{0'} x^{\mu'}(X^1/X^0, X^2/X^0, X^3/X^0, X^4/X^0) \tag{66}$$

It is now easy to verify that $c^{-1}\mathscr{G}c = \mathscr{G}$ also holds in $\mathscr{H}$. Thus $\mathscr{H}$ is the underlying group for the Einstein-Maxwell equation and this fact forms the basis of Veblen's projective theory of relativity. However, it can be shown that Veblen's projective theory of relativity is essentially *equivalent* to the above (Kaluza and Klein's) 5-dimensional formalism. The correspondence between the projective co-ordinates X^i and the x^i is given by

$$X^i = f^i(x^\mu)e^{x^0} \tag{67}$$

with arbitrary f^i. The inverse to (67) is

$$x^\mu = g^\mu(X^1/X^0, \ldots X^4/X^0) \tag{68}$$

$$x^0 = \log\{X^0 h(X^1/X^0, \ldots X^4/X^0)\} = \log G(X^k)$$

where G is a homogeneous function of degree one. The transformations (48) thus correspond exactly to the group $\mathscr{H}$ of (63).

Kaluza-Klein's theory or the equivalent projective theory of relativity can be generalized† by either enlarging the transformation group (48) or by letting γ_{00} be a variable. These generalizations have introduced certain novel features in 5-dimensional theories: (i) the constant of gravitation κ is no longer a constant; (ii) the motion of *uncharged* matter can create a magnetic field, an effect hypothesized heuristically by Blackett.‡ We now consider the other approach to the problem of unification, which is perhaps the most celebrated of all—namely, the non-symmetric theory of Einstein and Schrödinger.

Non-symmetric field theory. One obtains a fruitful generalization of Riemannian geometry by abandoning the demand of symmetry of the components of affine connection $\Gamma^\mu_{\alpha\beta}$. The parallel transfer of a vector in a non-symmetric affinely connected manifold is given by either of the two forms

$$\delta\xi^\mu_+ = -\Gamma^\mu_{\alpha\beta}\xi^\alpha\, dx^\beta \tag{69}$$

$$\delta\xi^\mu_- = -\Gamma^\mu_{\alpha\beta}\xi^\beta\, dx^\alpha \tag{70}$$

Consequently one can define two types of covariant derivatives

$$\xi^\mu_{+;\rho} = \xi^\mu_{,\rho} + \Gamma^\mu_{\sigma\rho}\xi^\sigma \tag{71}$$

$$\xi^\mu_{-;\rho} = \xi^\mu_{,\rho} + \Gamma^\mu_{\rho\sigma}\xi^\sigma \tag{72}$$

† Jordan, P. (1947) *Ann. Phys.* **1**, 219. Thiry, Y. (1950) Thèse, Paris.
‡ Blackett, P. M. S. (1948) Solvay Congress.

We denote the symmetric and antisymmetric parts of $\Gamma^{\mu}_{\alpha\beta}$ by $\Gamma^{\mu}_{\underline{\alpha\beta}}$ and $\Gamma^{\mu}_{\underset{\vee}{\alpha\beta}}$ respectively

$$\Gamma^{\mu}_{\alpha\beta} = \Gamma^{\mu}_{\underline{\alpha\beta}} + \Gamma^{\mu}_{\underset{\vee}{\alpha\beta}}; \qquad \Gamma^{\mu}_{\underline{\alpha\beta}} = \Gamma^{\mu}_{\underline{\beta\alpha}}; \qquad \Gamma^{\mu}_{\underset{\vee}{\alpha\beta}} = -\Gamma^{\mu}_{\underset{\vee}{\beta\alpha}} \tag{73}$$

Consider now an *arbitrary* second-order tensor $g_{\mu\lambda}$ and define $g^{\mu\lambda}$ by (assuming $g \equiv |g_{\mu\lambda}| \neq 0$)

$$g_{\mu\rho}g^{\mu\sigma} = g_{\rho\mu}g^{\sigma\mu} = \delta^{\sigma}_{\rho} \tag{74}$$

Let $g_{\underline{\mu\lambda}}$, $g^{\underline{\mu\lambda}}$ be the symmetric and $g_{\underset{\vee}{\mu\lambda}}$, $g^{\underset{\vee}{\mu\lambda}}$ the antisymmetric parts of $g_{\mu\lambda}$, $g^{\mu\lambda}$ respectively. We have

$$g_{\mu\lambda} = g_{\underline{\mu\lambda}} + g_{\underset{\vee}{\mu\lambda}} \tag{75}$$

$$g^{\mu\lambda} = g^{\underline{\mu\lambda}} + g^{\underset{\vee}{\mu\lambda}} \tag{76}$$

where

$$g_{\underline{\mu\lambda}} = g_{\underline{\lambda\mu}}, \qquad g_{\underset{\vee}{\mu\lambda}} = -g_{\underset{\vee}{\lambda\mu}} \quad \text{etc.} \tag{77}$$

According to (69) and (70) the covariant derivatives of $g_{\mu\nu}$, $g^{\mu\nu}$ are given by

$$g_{\underset{+}{\mu}\underset{-}{\nu};\rho} = g_{\mu\nu,\rho} - \Gamma^{\sigma}_{\mu\rho}g_{\sigma\nu} - \Gamma^{\sigma}_{\rho\nu}g_{\mu\sigma} \tag{78}$$

$$g^{\mu\nu}_{+-;\rho} = g^{\mu\nu}{}_{,\rho} + \Gamma^{\mu}_{\sigma\rho}g^{\sigma\nu} + \Gamma^{\nu}_{\rho\sigma}g^{\mu\sigma} \tag{79}$$

The covariant derivative of the tensor-density

$$\mathfrak{g}^{\mu\nu} \equiv \sqrt{(-g)}\, g^{\mu\nu} \tag{80}$$

is then

$$\mathfrak{g}^{\mu\nu}_{+-;\rho} = \mathfrak{g}^{\mu\nu}{}_{,\rho} + \Gamma^{\mu}_{\sigma\rho}\mathfrak{g}^{\sigma\nu} + \Gamma^{\nu}_{\rho\sigma}\mathfrak{g}^{\mu\sigma} - \mathfrak{g}^{\mu\nu}\Gamma^{\sigma}_{\underline{\rho\sigma}} \tag{81}$$

The parallel transfer (70) can also be written as

$$\delta\xi^{\underline{\mu}} = -\tilde{\Gamma}^{\mu}_{\alpha\beta}\xi^{\alpha}\, dx^{\beta} \tag{82}$$

where

$$\tilde{\Gamma}^{\mu}_{\alpha\beta} = \Gamma^{\mu}_{\beta\alpha}$$

We can thus define two curvature tensors in a non-symmetric manifold:

$$R^{\rho}_{\mu\nu\rho}(\Gamma) = \Gamma^{\rho}_{\mu\nu,\sigma} - \Gamma^{\rho}_{\mu\sigma,\nu} + \Gamma^{\lambda}_{\mu\nu}\Gamma^{\rho}_{\lambda\sigma} - \Gamma^{\lambda}_{\mu\sigma}\Gamma^{\rho}_{\lambda\nu} \tag{83}$$

$$R^{\rho}_{\mu\nu\sigma}(\tilde{\Gamma}) = \Gamma^{\rho}_{\nu\mu,\sigma} - \Gamma^{\rho}_{\sigma\mu,\nu} + \Gamma^{\lambda}_{\nu\mu}\Gamma^{\rho}_{\sigma\lambda} - \Gamma^{\lambda}_{\sigma\mu}\Gamma^{\rho}_{\nu\lambda} \tag{84}$$

Each of the curvature tensors gives rise to two contracted tensors, for example:

$$R_{\mu\nu}(\Gamma) \equiv R^{\rho}_{\mu\nu\rho}(\Gamma) = \Gamma^{\rho}_{\mu\nu,\rho} - \Gamma^{\rho}_{\mu\rho,\nu} + \Gamma^{\lambda}_{\mu\nu}\Gamma^{\rho}_{\lambda\rho} - \Gamma^{\lambda}_{\mu\rho}\Gamma^{\rho}_{\lambda\nu} \tag{85}$$

$$S_{\mu\nu}(\Gamma) \equiv R^{\rho}_{\rho\mu\nu}(\Gamma) = \Gamma^{\rho}_{\rho\mu,\nu} - \Gamma^{\rho}_{\rho\nu,\mu} \tag{86}$$

and

$$R_{\mu\nu}(\tilde{\Gamma}) \equiv R^{\rho}_{\mu\nu\rho}(\tilde{\Gamma}) = \Gamma^{\rho}_{\nu\mu,\rho} - \Gamma^{\rho}_{\rho\mu,\nu} + \Gamma^{\lambda}_{\nu\mu}\Gamma^{\rho}_{\rho\lambda} - \Gamma^{\lambda}_{\rho\mu}\Gamma^{\rho}_{\nu\lambda} \tag{87}$$

$$S_{\mu\nu}(\tilde{\Gamma}) \equiv R^{\rho}_{\rho\mu\nu}(\tilde{\Gamma}) = \Gamma^{\rho}_{\mu\rho,\nu} - \Gamma^{\rho}_{\nu\rho,\mu} \tag{88}$$

In a symmetric manifold (85) is same as (87) and both (86) and (88) vanish. Besides the above four second-order tensors there is another basic tensor in a non-symmetric manifold:

$$\Gamma_{\mu\nu} = \Gamma_\mu \Gamma_\nu \quad \text{where} \quad \Gamma_\mu \equiv \Gamma^\rho_{\underset{\vee}{\mu\rho}} \tag{89}$$

is called the torsion vector of the manifold.

We thus have an *embarras de choix* for forming a variational principle for the field-equations. Following Einstein, one can reduce the number of basic tensors by demanding that they be *transposition invariant*. Consider, for example, a general tensor $A_{\mu\lambda}(g_{\mu\lambda}, \Gamma^\mu_{\alpha\beta})$ which is composed of $g_{\mu\lambda}$ and $\Gamma^\mu_{\alpha\beta}$. Then $A_{\mu\lambda}(g, \Gamma)$ is said to be transposition invariant if

$$A_{\mu\lambda}(g, \Gamma) = A_{\lambda\mu}(\tilde{g}, \tilde{\Gamma}) \tag{90}$$

where again

$$\tilde{g}_{\mu\lambda} = g_{\lambda\mu}$$

None of (85)–(88) is transposition invariant; but any linear combination of the following tensors is an invariant under the combined transposition of Γ by $\tilde{\Gamma}$ and the index μ by λ.

$$\begin{aligned} A_{\mu\lambda} &= R_{\mu\lambda}(\Gamma) + R_{\lambda\mu}(\tilde{\Gamma}) \\ B_{\mu\lambda} &= S_{\mu\lambda}(\Gamma) + S_{\lambda\mu}(\tilde{\Gamma}) \\ \Gamma_{\mu\lambda} &= \Gamma_\mu \Gamma_\lambda \end{aligned}$$

We suppose now that the non-symmetric manifold is characterized also by a non-symmetric "metric" tensor $g_{\mu\lambda}$ and consider the following variational principle

$$\delta \int \mathscr{W}(\mathfrak{g}^{\mu\lambda}; \Gamma^\mu_{\alpha\beta}; \Gamma^\mu_{\alpha\beta,\gamma})\, dv = 0 \tag{91}$$

where the variables $\mathfrak{g}^{\mu\lambda}$ and $\Gamma^\mu_{\alpha\beta}$ are to be regarded as independent and the variations $\delta\mathfrak{g}^{\mu\lambda}$ and $\delta\Gamma^\mu_{\alpha\beta}$ vanish at the boundary. We can then write (91) as

$$\int \left[\mathscr{W}^{\alpha\beta}_\mu \,\delta\Gamma^\mu_{\alpha\beta} + w_{\mu\nu}\,\delta\mathfrak{g}^{\mu\nu}\right] dv = 0 \tag{92}$$

from which follows:

$$\mathscr{W}^{\alpha\beta}_\mu \equiv \delta\mathscr{W}/\delta\Gamma^\mu_{\alpha\beta} = 0 \tag{93}$$

$$w_{\mu\nu} = \delta\mathscr{W}/\delta\mathfrak{g}^{\mu\nu} = 0 \tag{94}$$

If we take

$$\mathscr{W} = \mathfrak{g}^{\mu\nu} R_{\mu\nu}(\Gamma) \tag{95}$$

then

$$w_{\mu\nu} = \delta \mathscr{W} / \delta \mathfrak{g}^{\mu\nu} = R_{\mu\nu}(\Gamma)$$

so that (94) becomes

$$R_{\mu\nu}(\Gamma) = 0 \tag{96}$$

If, on the other hand, we take

$$\mathscr{W} = \mathfrak{g}^{\mu\nu} R_{\mu\nu}(\Gamma) - 2\lambda\sqrt{-g} \tag{97}$$

where λ is a constant, we get instead

$$R_{\mu\nu}(\Gamma) - \lambda g_{\mu\nu} = 0 \tag{98}$$

Instead of considering $\mathfrak{g}^{\mu\nu}$ and $\Gamma^{\mu}_{\alpha\beta}$ as independent variables we can also proceed from an action function $\mathscr{W}$ depending *only* on $\Gamma^{\mu}_{\alpha\beta}$ and $\Gamma^{\mu}_{\alpha\beta,\gamma}$

$$\delta \int \mathscr{W}(\Gamma^{\mu}_{\alpha\beta}; \Gamma^{\mu}_{\alpha\beta,\gamma})\, dv = 0 \tag{99}$$

or

$$\int \mathscr{W}^{\alpha\beta}_{\mu}\, \delta\Gamma^{\mu}_{\alpha\beta}\, dv = 0$$

so that the Euler-Lagrange equations are only

$$\mathscr{W}^{\alpha\beta}_{\mu} \equiv \frac{\delta \mathscr{W}}{\delta \Gamma^{\mu}_{\alpha\beta}} \equiv \frac{\partial \mathscr{W}}{\partial \Gamma^{\mu}_{\alpha\beta}} - \frac{\partial}{\partial x^{\gamma}}\left(\frac{\partial \mathscr{W}}{\partial \Gamma^{\mu}_{\alpha\beta,\gamma}}\right) = 0 \tag{100}$$

We obtain a second system of equations by now *defining* the tensor density $\mathfrak{g}^{\mu\nu}$ by

$$\mathfrak{g}^{\mu\nu} = \frac{\delta \mathscr{W}}{\delta R_{\mu\nu}(\Gamma)} \equiv \frac{\partial \mathscr{W}}{\partial R_{\mu\nu}(\Gamma)} \tag{101}$$

If we now take

$$\mathscr{W} = \frac{2}{\lambda}\sqrt{(-|R_{\mu\nu}(\Gamma)|)} \tag{102}$$

then (101) is equivalent to (98).

Equation (93) for $\mathscr{W} = \mathfrak{g}^{\mu\nu} R_{\mu\nu}(\Gamma)$ becomes

$$\mathscr{W}^{\alpha\beta}_{\mu} \equiv \mathfrak{g}^{\lambda\nu} \frac{\partial R_{\lambda\nu}(\Gamma)}{\partial \Gamma^{\mu}_{\alpha\beta}} - \frac{\partial}{\partial x^{\gamma}}\left(\mathfrak{g}^{\lambda\nu} \frac{\partial R_{\lambda\nu}(\Gamma)}{\partial \Gamma^{\mu}_{\alpha\beta,\gamma}}\right) = 0$$

or

$$\mathfrak{g}^{\alpha\beta}_{,\mu} - \delta^{\beta}_{\mu}\mathfrak{g}^{\alpha\nu}_{,\nu} - \mathfrak{g}^{\alpha\beta}\Gamma^{\rho}_{\mu\rho} - \delta^{\beta}_{\mu}\mathfrak{g}^{\rho\sigma}\Gamma^{\alpha}_{\rho\sigma} + \mathfrak{g}^{\alpha\rho}\Gamma^{\beta}_{\mu\rho} + \mathfrak{g}^{\rho\beta}\Gamma^{\alpha}_{\rho\mu} = 0 \tag{103}$$

Contracting first β and μ, then α and μ we get

$$-3\mathfrak{g}^{\alpha\mu}_{,\mu} - 2\mathfrak{g}^{\alpha\mu}\Gamma^{\rho}_{\underset{\vee}{\mu\rho}} - 3\mathfrak{g}^{\sigma\rho}\Gamma^{\alpha}_{\sigma\rho} = 0 \tag{104}$$

$$\mathfrak{g}^{\underset{\vee}{\alpha\beta}}{}_{,\alpha} = 0 \tag{105}$$

where $\underset{\vee}{\mathfrak{g}}^{\alpha\beta} = \sqrt{(-g)}\, \underset{\vee}{g}^{\alpha\beta}$. From (104) and (103) we get

$$\mathfrak{g}^{\underset{+}{\alpha}\underset{-}{\beta}}{}_{;\mu} + \tfrac{2}{3}\delta^{\beta}_{\mu}\mathfrak{g}^{\alpha\sigma}\Gamma_{\sigma} - \mathfrak{g}^{\alpha\beta}\Gamma_{\mu} = 0 \tag{106}$$

The field equations (93) and (94) for $\mathscr{W} = \mathfrak{g}^{\mu\nu}R_{\mu\nu}(\Gamma)$ are thus equivalent to (96), (105) and (106) which are grouped together here for reference.

$$\left.\begin{aligned} \mathfrak{g}^{\underset{+}{\alpha}\underset{-}{\beta}}{}_{;\mu} &= -\tfrac{2}{3}\delta^{\beta}_{\mu}\mathfrak{g}^{\alpha\sigma}\Gamma_{\sigma} + \mathfrak{g}^{\alpha\beta}\Gamma_{\mu} \\ \underset{\vee}{\mathfrak{g}}^{\alpha\beta}{}_{,\mu} &= 0 \\ R_{\mu\nu}(\Gamma) &= 0 \end{aligned}\right\} \tag{A}$$

The equations (A) which follow strictly from the variational principle (91) could be simplified by imposing the additional condition $\Gamma_{\sigma} = 0$. We then obtain another set of field-equations which are *not* deducible from (91)

$$\left.\begin{aligned} \mathfrak{g}^{\underset{+}{\alpha}\underset{-}{\beta}}{}_{;\mu} &= 0, \qquad & \Gamma_{\sigma} &= 0 \\ \underset{\vee}{\mathfrak{g}}^{\alpha\beta}{}_{,\alpha} &= 0, \qquad & R_{\mu\nu}(\Gamma) &= 0 \end{aligned}\right\} \tag{B}$$

However, Einstein and Kaufman† have demonstrated that the equations (B) are unduly restrictive from the physical standpoint. Equations (A) can be expressed in a simpler form by introducing a new affine connection whose torsion vector vanishes. Let

$$L^{\mu}_{\alpha\beta} = \Gamma^{\mu}_{\alpha\beta} + \tfrac{2}{3}\,\delta^{\mu}_{\alpha}\Gamma_{\beta} \tag{107}$$

then

$$L_{\mu} \equiv \underset{\vee}{L^{\rho}_{\mu\rho}} = 0 \tag{108}$$

(106) is then equivalent to

$$\mathfrak{g}^{\underset{+}{\alpha}\underset{-}{\beta}}{}_{|\mu} \equiv \mathfrak{g}^{\alpha\beta}{}_{,\mu} + \mathfrak{g}^{\alpha\sigma}L^{\beta}_{\mu\sigma} + \mathfrak{g}^{\sigma\beta}L^{\alpha}_{\sigma\mu} - \mathfrak{g}^{\alpha\beta}L^{\rho}_{\mu\rho} = 0 \tag{109}$$

The left-hand side is a covariant derivative with respect to the new connection $L^{\mu}_{\alpha\beta}$.

Written in terms of $L^{\mu}_{\alpha\beta}$ (96) becomes

$$R_{\mu\nu}(\Gamma) \equiv R_{\mu\nu}(L) - \tfrac{2}{3}(\Gamma_{\nu,\mu} - \Gamma_{\mu,\nu}) = 0 \tag{110}$$

where $R_{\mu\nu}(L)$ is the contracted curvature tensor relative to the new connection. (110) can be split into a symmetric and antisymmetric part

$$R_{\underline{\mu\nu}}(L) = 0 \tag{111}$$

$$R_{\underset{\vee}{\mu\nu}}(L) + \tfrac{2}{3}(\Gamma_{\nu,\mu} - \Gamma_{\mu,\nu}) = 0 \tag{112}$$

Equation (112) implies and is, in fact, equivalent to

$$R_{\underset{\vee}{\mu\nu}}(L)_{,\lambda} + R_{\underset{\vee}{\nu\lambda}}(L)_{,\mu} + R_{\underset{\vee}{\lambda\mu}}(L)_{,\nu} = 0 \tag{113}$$

† Einstein, A. and Kaufman, B. (1952) "Louis de Broglie, Physicien et Penseur", p. 321, Albin Michel, Paris.

The set (A) relative to the connection (107) is thus equivalent to

$$\left.\begin{array}{ll} \mathfrak{g}^{\underset{+}{\alpha}\underset{-}{\beta}}{}_{|\mu} = 0; & \mathfrak{g}^{\underset{\vee}{\alpha\beta}}{}_{,\alpha} = 0 \\ R_{\underline{\mu\nu}}(L) = 0; & R_{\underset{\vee}{\mu\nu}}(L)_{,\lambda} + R_{\underset{\vee}{\nu\lambda}}(L)_{,\mu} + R_{\underset{\vee}{\lambda\mu}}(L)_{,\nu} = 0 \end{array}\right\} \qquad (\bar{\mathrm{A}})$$

The system $(\bar{\mathrm{A}})$ has the property that if $(g_{\lambda\mu}, L^{\lambda}_{\alpha\beta})$ is one of its solutions, then so also is $(\tilde{g}_{\lambda\mu}, \tilde{L}^{\lambda}_{\alpha\beta})$. The invariance of the action integral in (91) under arbitrary co-ordinate transformations furnishes us with several identities which may be identified with equations of conservation.

Consider an infinitesimal change in the co-ordinates

$$x^{\mu} \to x^{\mu'} = x^{\mu} + \delta x^{\mu}, \qquad \delta x^{\mu} = 0 \text{ at the boundary}$$

We write the variation (91) due to the above change in the co-ordinates as (using (95))

$$\int [(\delta R_{\mu\nu}(\Gamma))\mathfrak{g}^{\mu\nu} + R_{\mu\nu}(\Gamma)\,\delta\mathfrak{g}^{\mu\nu}]\,dv = 0 \tag{114}$$

The first integral vanishes due to (103) and (a similar situation arises in the symmetric case where the first integral vanishes in view of $\mathfrak{g}^{\mu\lambda}{}_{;\alpha} = 0$) we are left with

$$\int R_{\mu\nu}(\Gamma)\,\delta\mathfrak{g}^{\mu\nu}\,dv = 0 \tag{115}$$

Now the change in $g^{\mu\nu}$ due to the change of co-ordinates is

$$\delta g^{\mu\nu} = -g^{\mu\nu}{}_{,\rho}\,\delta x^{\rho} + g^{\mu\rho}(\delta x^{\nu})_{,\rho} + g^{\rho\nu}(\delta x^{\mu})_{,\rho} \tag{116}$$

Therefore

$$\begin{aligned} -R_{\mu\nu}(\Gamma)\,\delta\mathfrak{g}^{\mu\nu} = {} & \left[R_{\mu\nu}(\Gamma)\mathfrak{g}^{\mu\nu}\,\delta x^{\rho} - (R_{\mu\nu}(\Gamma)\mathfrak{g}^{\rho\nu} + R_{\nu\mu}(\Gamma)\mathfrak{g}^{\nu\rho})\,\delta x^{\mu}\right]_{,\rho} \\ & + \left[(R_{\mu\nu}(\Gamma)\mathfrak{g}^{\rho\nu} + R_{\nu\mu}(\Gamma)\mathfrak{g}^{\nu\rho})_{,\rho} - \delta^{\rho}_{\mu}\mathfrak{g}^{\alpha\beta}R_{\alpha\beta}(\Gamma)_{,\rho}\right]\delta x^{\mu} \end{aligned}$$

The first term on the right is a divergence and therefore does not contribute anything to (115). For arbitrary δx^{μ} we must then have the identities

$$(R_{\mu\nu}(\Gamma)\mathfrak{g}^{\rho\nu} + R_{\nu\mu}(\Gamma)\mathfrak{g}^{\nu\rho})_{,\rho} - \delta^{\rho}_{\mu}\mathfrak{g}^{\alpha\beta}R_{\alpha\beta}(\Gamma)_{,\rho} = 0 \tag{117}$$

Let
$$H^{\rho}_{\mu} \equiv \tfrac{1}{2}(R_{\mu\nu}(\Gamma)g^{\rho\nu} + R_{\nu\mu}(\Gamma)g^{\nu\rho}) - \tfrac{1}{2}\,\delta^{\rho}_{\mu}R(\Gamma) \tag{118}$$

with
$$R(\Gamma) = g^{\mu\nu}R_{\mu\nu}(\Gamma)$$

and $\mathscr{H}^{\rho}_{\mu} = \sqrt{(-g)}\,H^{\rho}_{\mu}$. Then (117) can be written in the form

$$\mathscr{H}^{\rho}_{\mu,\rho} + \tfrac{1}{2}R_{\alpha\beta}(\Gamma)\mathfrak{g}^{\alpha\beta}{}_{,\mu} = 0 \tag{119}$$

If we want to express (119) in terms of the new connection $L^{\mu}_{\alpha\beta}$ we must use (110), that is,

$$R_{\mu\nu}(\Gamma) = R_{\mu\nu}(L) - \tfrac{2}{3}(\Gamma_{\nu,\mu} - \Gamma_{\mu,\nu})$$

so that $$\mathscr{H}^{\rho}_{\mu} = \mathscr{K}^{\rho}_{\mu} - \tfrac{2}{3}(\psi_{\mu\nu}\mathscr{g}^{\rho\nu} - \tfrac{1}{2}\,\delta^{\rho}_{\mu}\psi_{\alpha\beta}\mathscr{g}^{\alpha\beta})$$

where we have put

$$\psi_{\mu\nu} = \Gamma_{\nu,\mu} - \Gamma_{\mu,\nu}$$
$$K^{\rho}_{\mu} = \tfrac{1}{2}(R_{\mu\nu}(L)g^{\rho\nu} + R_{\mu\nu}(L)g^{\nu\rho}) - \tfrac{1}{2}\,\delta^{\rho}_{\mu}R(L)$$
$$\mathscr{K}^{\rho}_{\mu} = \sqrt{(-g)}K^{\rho}_{\mu}$$

(119) thus becomes in view of (105)

$$\mathscr{K}^{\rho}_{\mu,\rho} + \tfrac{1}{2}R_{\alpha\beta}(L)\mathscr{g}^{\alpha\beta}{}_{,\mu} = 0 \tag{120}$$

The identities thus have the same form relative to the connections Γ or L. We shall come back to the identities later.

To proceed further one must now solve (109) for the affine connection $\Gamma^{\mu}_{\alpha\beta}$ or $L^{\mu}_{\alpha\beta}$ in terms of the "metric" tensor $g_{\mu\lambda}$. In the Riemannian case this corresponds to solving $g_{\mu\lambda;\alpha} = 0$ for the *symmetric* $\Gamma^{\mu}_{\alpha\beta}$ in terms of the *symmetric* $g_{\mu\lambda}$; the unique solution in the symmetric case is of course $\Gamma^{\mu}_{\alpha\beta} = \left\{{\mu \atop \alpha\beta}\right\}$.

Equation (109), that is, $\mathscr{g}^{\underset{+}{\mu}\underset{-}{\nu}}{}_{|\rho} = 0$ is equivalent to either of the following

$$g^{\underset{+}{\mu}\underset{-}{\nu}}{}_{|\rho} \equiv g^{\mu\nu}{}_{,\rho} + L^{\mu}_{\sigma\rho}g^{\sigma\nu} + L^{\nu}_{\rho\sigma}g^{\mu\sigma} = 0 \tag{121}$$

$$g_{\underset{+}{\mu}\underset{-}{\nu}|\rho} \equiv g_{\mu\nu,\rho} - L^{\sigma}_{\mu\rho}g_{\sigma\nu} - L^{\sigma}_{\rho\nu}g_{\mu\sigma} = 0 \tag{122}$$

Multiplying (122) by $g^{\mu\nu}$ and summing we get

$$g^{\mu\nu}g_{\mu\nu,\rho} - L^{\mu}_{\mu\rho} - L^{\mu}_{\rho\mu} = 0$$

or

$$L^{\mu}_{\underline{\rho\mu}} = (\log\sqrt{-g})_{,\rho} \tag{123}$$

On the other hand, from (121) and (123), we get

$$0 = \tfrac{1}{2}(g^{\underset{+}{\mu}\underset{-}{\rho}}{}_{|\rho} - g^{\underset{+}{\rho}\underset{-}{\mu}}{}_{|\rho}) = g^{\underset{\vee}{\mu\rho}}{}_{,\rho} + L^{\rho}_{\underline{\sigma\rho}}g^{\underset{\vee}{\mu\sigma}} - L^{\rho}_{\underset{\vee}{\sigma\rho}}g^{\underline{\mu\sigma}}$$

or

$$\left(\frac{1}{\sqrt{-g}}\mathscr{g}^{\underset{\vee}{\mu\rho}}{}_{,\rho}\right) - L^{\rho}_{\underset{\vee}{\sigma\rho}}\mathscr{g}^{\underline{\mu\sigma}} = 0 \tag{124}$$

Therefore, (105) is a consequence of (108) and (109). It suffices therefore to solve (122) for the connection $L^{\mu}_{\alpha\beta}$ in terms of $g_{\mu\lambda}$ with the stipulation that $L_{\rho} \equiv L^{\rho}_{\underset{\vee}{\rho\sigma}} = 0$. Equation (122) breaks up into the following symmetric and antisymmetric parts in μ and ν:

$$g_{\underline{\mu\nu},\rho} - (L^{\sigma}_{\underline{\mu\rho}}g_{\underline{\sigma\nu}} + L^{\sigma}_{\underline{\nu\rho}}g_{\underline{\sigma\mu}}) - (L^{\sigma}_{\underset{\vee}{\mu\rho}}g_{\underset{\vee}{\sigma\nu}} + L^{\sigma}_{\underset{\vee}{\nu\rho}}g_{\underset{\vee}{\sigma\mu}}) = 0 \tag{125}$$

$$g_{\underset{\vee}{\mu\nu},\rho} - (L^{\sigma}_{\underline{\mu\rho}}g_{\underset{\vee}{\sigma\nu}} + L^{\sigma}_{\underline{\nu\rho}}g_{\underset{\vee}{\sigma\mu}}) - (L^{\sigma}_{\underset{\vee}{\mu\rho}}g_{\underline{\sigma\nu}} + L^{\sigma}_{\underset{\vee}{\nu\rho}}g_{\underline{\sigma\mu}}) = 0 \tag{126}$$

The symmetric and antisymmetric parts of L can then be written in the following form:

$$L^{\mu}_{\alpha\beta} = L^{\mu}_{\underline{\alpha\beta}} + L^{\mu}_{\underset{\vee}{\alpha\beta}} = \left\{ {\mu \atop \alpha\beta} \right\}_{g_{\underline{\nu\lambda}}} + u^{\mu}_{\alpha\beta} + L^{\mu}_{\underset{\vee}{\alpha\beta}} \tag{127}$$

where

$$u^{\sigma}_{\mu\nu} g_{\underline{\rho\sigma}} = L^{\sigma}_{\underset{\vee}{\mu\rho}} g_{\underset{\vee}{\sigma\nu}} + L^{\sigma}_{\underset{\vee}{\nu\rho}} g_{\underset{\vee}{\sigma\mu}} \tag{128}$$

and

$$L^{\sigma}_{\underset{\vee}{\mu\nu}} g_{\underline{\rho\sigma}} = -\tfrac{1}{2}(g_{\underset{\vee}{\mu\nu},\rho} + g_{\underset{\vee}{\nu\rho},\mu} + g_{\underset{\vee}{\rho\mu},\nu}) + g_{\underset{\vee}{\mu\nu};\rho} - (u^{\sigma}_{\mu\rho} g_{\underset{\vee}{\sigma\nu}} - u^{\sigma}_{\nu\rho} g_{\underset{\vee}{\sigma\mu}}) \tag{129}$$

Here $\left\{ {\mu \atop \alpha\beta} \right\}_{g_{\underline{\nu\lambda}}}$ denotes the Christoffel symbols formed with $g_{\underline{\nu\lambda}}$, and $g_{\underset{\vee}{\mu\nu};\rho}$ the covariant derivative of $g_{\underset{\vee}{\mu\nu}}$ relative to $\left\{ {\mu \atop \alpha\beta} \right\}_{g_{\underline{\lambda\nu}}}$. Thus, apart from $\left\{ {\mu \atop \alpha\beta} \right\}_{g_{\underline{\lambda\nu}}}$, the symmetric part of $L^{\mu}_{\alpha\beta}$ consists of $u^{\mu}_{\alpha\beta}$, which can be expressed, according to (128), in terms of the antisymmetric part $L^{\mu}_{\underset{\vee}{\alpha\beta}}$ which itself, when $u^{\sigma}_{\mu\nu}$ is substituted from (128) into (129), is finally determined according to (129) in terms of the basic fields $g_{\underline{\mu\lambda}}$ and $g_{\underset{\vee}{\mu\lambda}}$. We forego an explicit solution for $L^{\mu}_{\alpha\beta}$.

$L^{\mu}_{\alpha\beta}$ has a unique solution† provided

$$\underline{g} \equiv |g_{\underline{\mu\lambda}}| \neq 0 \tag{130}$$

and

$$(a^2 + b^2) \neq 0 \tag{131}$$

where

$$a \equiv 2 - \frac{g}{\underline{g}} + \frac{6\underset{\vee}{g}}{\underline{g}}, \qquad b \equiv \frac{2\sqrt{\underset{\vee}{g}}}{\sqrt{-\underline{g}}}\left(3 - \frac{g}{\underline{g}} + \frac{\underset{\vee}{g}}{\underline{g}}\right) \tag{132}$$

$$\underset{\vee}{g} \equiv |g_{\underset{\vee}{\mu\lambda}}|$$

assuming always $\underline{g} \neq 0$. If $\underset{\vee}{g} = 0$ then $b = 0$ and therefore the condition (131) implies $g \neq 2\underline{g}$. On the other hand the integrability condition for (121), regarded as a differential equation for $g^{\mu\nu}$, is

$$g^{\mu\nu}{}_{,\sigma,\rho} - g^{\mu\nu}{}_{,\rho,\sigma} = 0 \tag{133}$$

Substituting $g^{\mu\nu}{}_{,\sigma}$ from (121) in (133) we get

$$R_{\lambda\sigma\rho}{}^{\mu}(L) g^{\lambda\nu} + R_{\lambda\sigma\rho}{}^{\nu}(\tilde{L}) g^{\mu\lambda} = 0 \tag{134}$$

Contracting (134) with $g_{\mu\nu}$ one obtains

$$R^{\mu}_{\mu\sigma\rho}(L) + R^{\mu}_{\mu\sigma\rho}(\tilde{L}) = 0$$

or

$$L^{\mu}_{\underline{\mu\sigma},\rho} - L^{\mu}_{\underline{\mu\rho},\sigma} = 0 \tag{135}$$

† Hlavaty, V. (1952) *Proc. natn. Acad. Sci. U.S.A.* **38**, no. 5, 415; no. 12, 1052. Tonnelat, M. A. (1955) *J. Phys. Rad., Paris* **16**, 21.

(135) is identically satisfied by (123). And contracting (134) first with respect to ρ and μ and then σ and ν we get

$$R^{\mu}_{\lambda\nu\mu}(L)g^{\lambda\nu}+R^{\nu}_{\lambda\nu\mu}(\tilde{L})g^{\mu\lambda} = 0$$

or

$$\{R_{\lambda\nu}(L)-R_{\nu\lambda}(\tilde{L})\}g^{\lambda\nu} = 0$$

or

$$g^{\lambda\nu}\{(L_{\nu,\lambda}-L_{\lambda,\nu})-2L^{\sigma}_{\check{\lambda\nu}}L_{\sigma}\} = 0 \tag{136}$$

which is also identically satisfied by (108).

The field-equations ($\bar{\text{A}}$) therefore become (in view of (127)) together with (105)

$$R_{\underline{\mu\nu}}(L) \equiv R_{\mu\nu}(\{{}^{\lambda}_{\alpha\beta}\})+u^{\rho}_{\mu\nu;\rho}-\tfrac{1}{2}(\log g)_{;\nu;\mu}+u^{\rho}_{\mu\nu}u^{\lambda}_{\rho\lambda} -(u^{\lambda}_{\mu\rho}u^{\rho}_{\lambda\nu}+L^{\lambda}_{\check{\mu\rho}}L^{\rho}_{\check{\lambda\nu}}) = 0 \tag{137}$$

$$R_{\check{\mu\nu}}(L) \equiv L^{\rho}_{\check{\mu\nu};\rho}+L^{\rho}_{\check{\mu\nu}}u^{\lambda}_{\rho\lambda}-(u^{\lambda}_{\mu\rho}L^{\rho}_{\check{\lambda\nu}}+u^{\rho}_{\lambda\nu}L^{\lambda}_{\check{\mu\rho}}) \tag{138}$$

$$R_{\check{\mu\nu},\alpha}(L)+R_{\check{\nu\alpha},\mu}(L)+R_{\check{\alpha\mu},\nu}(L) = 0 \tag{139}$$

where we have, for convenience, put $\{{}^{\lambda}_{\alpha\beta}\}_{g_{\underline{\mu\nu}}} \equiv \{{}^{\lambda}_{\alpha\beta}\}$ and we now suppose that $u^{\rho}_{\mu\nu}$ and $L^{\rho}_{\check{\mu\nu}}$ are expressed in terms of the fundamental fields $g_{\underline{\mu\lambda}}$ and $g_{\check{\mu\lambda}}$.

The above equations, being unusually complicated, have been subjected mainly to the following two procedures: (i) search for rigorous solutions in simple special cases, for example, static, spherically symmetric; and (ii) weak-field approximations of various orders. We consider briefly the latter first.

Let $\eta_{\mu\nu}$ be the Minkowskian values of the symmetric metric tensor, i.e., in Cartesian co-ordinates $\eta_{\mu\nu} = (-1, -1, -1, +1)$. Assume then that $g_{\check{\mu\nu}}$ and $g_{\underline{\mu\nu}}$ can be expanded in terms of a small parameter ϵ as follows:

$$g_{\underline{\mu\nu}} = \eta_{\mu\nu}+\epsilon \underset{1}{g_{\underline{\mu\nu}}}+\epsilon^{2}\underset{2}{g_{\underline{\mu\nu}}}+\cdots \tag{140}$$

$$g_{\check{\mu\nu}} = \epsilon \underset{1}{g_{\check{\mu\nu}}}+\epsilon^{2}\underset{2}{g_{\check{\mu\nu}}}+\cdots \tag{141}$$

Consequently we have also for the affine connection

$$\{{}^{\mu}_{\alpha\beta}\} = \epsilon\underset{1}{\{{}^{\mu}_{\alpha\beta}\}}+\epsilon^{2}\underset{2}{\{{}^{\mu}_{\alpha\beta}\}}+\cdots \tag{142}$$

$$L^{\mu}_{\check{\alpha\beta}} = \epsilon \underset{1}{L^{\mu}_{\check{\alpha\beta}}}+\epsilon^{2}\underset{2}{L^{\mu}_{\check{\alpha\beta}}}+\cdots \tag{143}$$

$$u^{\mu}_{\alpha\beta} = \epsilon^{2}\underset{2}{u^{\mu}_{\alpha\beta}}+\epsilon^{3}\underset{3}{u^{\mu}_{\alpha\beta}}+\cdots \tag{144}$$

The field-equations

$$\left.\begin{aligned} g^{\underset{\vee}{\alpha\beta}}{}_{,\alpha} &= 0 \\ R_{\underline{\mu\nu}}(L) &= 0 \\ R_{\underset{\vee}{\mu\nu},\lambda}(L)+R_{\underset{\vee}{\nu\lambda},\mu}(L)+R_{\underset{\vee}{\lambda\mu},\nu}(L) &= 0 \end{aligned}\right\} \tag{$\bar{\mathrm{A}}$}$$

become in *first order*

$$\eta^{\rho\nu}\underset{1}{g}_{\underset{\vee}{\mu\nu},\rho} = 0 \tag{145}$$

$$\eta^{\rho\nu}\{\eta_{\nu\lambda}\underset{1}{L}{}^{\lambda}_{\underset{\vee}{\mu\sigma}}\}_{,\rho,\tau} + \text{cyclic permutation of the indices } \tau,\ \mu,\ \sigma = 0 \tag{146}$$

$$\tfrac{1}{2}\eta^{\rho\sigma}(\underset{1}{g}_{\underline{\nu\sigma},\mu,\rho} - \underset{1}{g}_{\underline{\mu\nu},\sigma,\rho} - \underset{1}{g}_{\underline{\rho\sigma},\mu,\nu} + \underset{1}{g}_{\underline{\mu\rho},\sigma,\nu}) = 0 \tag{147}$$

and in the *second order*

$$\eta^{\rho\nu}\underset{2}{g}_{\underset{\vee}{\mu\nu},\rho} + \eta^{\rho\nu}\underset{1}{g}_{\underset{\vee}{\mu\nu}}(\sqrt{-\underset{1}{g}})_{,\rho} - \eta^{\sigma\lambda}\eta^{\rho\nu}\underset{1}{g}_{\underset{\vee}{\sigma\nu}}\underset{1}{g}_{\underline{\mu\lambda},\rho} - \eta^{\rho\nu}\eta^{\sigma\lambda}(\underset{1}{g}_{\underline{\nu\lambda}}\underset{1}{g}_{\underset{\vee}{\mu\sigma}})_{,\rho} = 0 \tag{148}$$

$$\Big[\underset{2}{L}{}^{\rho}_{\underset{\vee}{\mu\nu},\rho} + \underset{1}{\left\{{\rho \atop \sigma\rho}\right\}}\underset{1}{L}{}^{\sigma}{}_{\underset{\vee}{\mu\nu}} - \underset{1}{\left\{{\lambda \atop \mu\rho}\right\}}\underset{1}{L}{}^{\rho}_{\underset{\vee}{\lambda\nu}} - \underset{1}{\left\{{\lambda \atop \nu\rho}\right\}}\underset{1}{L}{}^{\rho}_{\underset{\vee}{\mu\lambda}}\Big]_{,\tau} + \text{cyclic permutation of } \tau,\ \mu,\ \nu = 0 \tag{149}$$

$$\underset{2}{R}_{\mu\nu}(\{{\lambda \atop \alpha\beta}\}) + \underset{2}{u}{}^{\rho}_{\mu\nu,\rho} - \tfrac{1}{2}\left\{\log\left(\frac{g}{\underset{2}{\underline{g}}}\right)\right\}_{,\nu,\mu} - \underset{1}{L}{}^{\lambda}_{\underset{\vee}{\mu\rho}}\underset{1}{L}{}^{\rho}_{\underset{\vee}{\lambda\nu}} = 0 \tag{150}$$

By considering corresponding equations for (B) Einstein and Kaufman† have shown that the set (B) is inadmissible from the physical point of view. In fact, the corresponding equations for (B) in the first-order approximation become together with (147)

$$\eta^{\rho\nu}\underset{1}{g}_{\underset{\vee}{\mu\nu},\rho} = 0 \tag{151}$$

$$\eta^{\rho\sigma}\underset{1}{g}_{\underset{\vee}{\mu\nu},\sigma,\rho} = 0 \tag{152}$$

On the other hand the following condition must be satisfied for the existence of a solution of the second-order equations

$$\underset{1}{g}_{\underset{\vee}{\mu\nu},\rho}(\underset{1}{g}_{\underline{\sigma\mu},\rho,\nu} + \underset{1}{g}_{\sigma\nu,\mu,\sigma}) = 0 \tag{153}$$

In the first order, therefore, the symmetric and antisymmetric components $\underset{1}{g}_{\underline{\mu\nu}}$, $\underset{1}{g}_{\underset{\vee}{\mu\nu}}$ satisfy independent equations and can thus be arbitrarily superimposed. In the second order, however, the condition (153) can be shown to be inadmissibly restrictive.

† See footnote, p. 93.

We now consider rigorous solutions of the field-equations ($\bar{\text{A}}$) under the simplifying assumption that the field $g_{\mu\nu}$ possesses spherical symmetry. In general relativity, where $g_{\mu\nu}$ is symmetric, a spherically symmetric metric in a suitable co-ordinate system (r, θ, ϕ, t) has the following form:

$$ds^2 = -\alpha\, dr^2 - \beta(d\theta^2 + \sin^2\theta\, d\phi^2) + \sigma\, dt^2$$

with

$$\alpha \equiv \alpha(r, t), \qquad \beta \equiv \beta(r, t), \qquad \sigma \equiv \sigma(r, t) \tag{154}$$

The well-known Schwarzschild solution of the gravitational field-equations $R_{\mu\nu} = 0$, in the *static*, spherically symmetric case is given by (see Part 1, equation (77)):

$$\beta = r^2, \qquad \sigma = \frac{1}{\alpha} = 1 - \frac{2m}{r} \tag{155}$$

In presence of an electromagnetic field the corresponding solution of the Einstein-Maxwell equations (1) is†

$$\beta = r^2, \qquad \sigma = \frac{1}{\alpha} = 1 - \frac{2m}{r} + \frac{4\pi e^2}{r^2} \tag{156}$$

where e, m can be interpreted as the charge and mass of the particle generating the gravitational and electrostatic field.

The form of a non-symmetric spherically symmetric $g_{\mu\nu}$ as proposed by Papapetrou‡ is

$$g_{\mu\nu} = \begin{pmatrix} -\alpha & 0 & 0 & w \\ 0 & -\beta & u\sin\theta & 0 \\ 0 & -u\sin\theta & -\beta\sin^2\theta & 0 \\ -w & 0 & 0 & \sigma \end{pmatrix} \tag{157}$$

where α, β, σ, u, w are functions of r and t. If one now computes the connection parameters $L^{\mu}_{\alpha\beta}$ from (127)–(129) one gets§

$$L^1_{\underline{11}} = \frac{\alpha'}{2\alpha} \qquad\qquad L^4_{\underline{44}} = \frac{\dot{\sigma}}{2\sigma}$$

$$L^1_{\underline{22}} = L^1_{\underline{33}}/\sin^2\theta = \frac{uB_1 - A_1}{2\alpha} \qquad\qquad L^4_{\underline{22}} = L^4_{\underline{33}}/\sin^2\theta = \frac{\beta A_4 - uB_4}{2\sigma}$$

† Nordstrom, G. (1918) *Verh. Ned. Akad. Wet.* **26**, 1201. Reissner, H. (1916) *Ann. Phys.* **50**, 106.
‡ Papapetrou, A. (1948) *Proc. R. Ir. Acad.* **52A**, 69.
§ Tonnelat, M. A. (1954) *C. R. Acad. Sci., Paris* **239**, 231.

$$L^1_{\underline{44}} = \frac{\sigma}{2\alpha}\left\{\log\sigma\left(1-\frac{w^2}{\alpha\sigma}\right)^2\right\}' \qquad L^4_{\underline{11}} = \frac{\alpha}{2\sigma}\overset{\cdot}{\Big\{}\log\alpha\left(1-\frac{w^2}{\alpha\sigma}\right)^2\Big\}$$

$$L^4_{\underline{14}} = \frac{1}{2}\left\{\log\sigma\left(1-\frac{w^2}{\alpha\sigma}\right)\right\}' \qquad L^1_{\underline{14}} = \frac{1}{2}\overset{\cdot}{\Big\{}\log\alpha\left(1-\frac{w^2}{\alpha\sigma}\right)\Big\}$$

$$L^2_{\underline{33}} = -\sin\theta\cos\theta \qquad L^3_{\underline{23}} = \frac{1}{\tan\theta} \tag{158}$$

$$L^2_{\underline{12}} = L^3_{\underline{31}} = \frac{A_1}{2} \qquad L^2_{\underline{24}} = L^3_{\underline{34}} = \frac{A_4}{2}$$

$$L^2_{\underline{34}} = -L^3_{\underline{24}}\sin^2\theta = \frac{wB_1}{2\alpha}\sin\theta \qquad L^2_{\underline{31}} = -L^3_{\underline{12}}\sin^2\theta = \frac{wB_4}{2\sigma}\sin\theta$$

$$L^1_{\underset{\vee}{23}} = \frac{\beta B_1+uA_1}{2\alpha}\sin\theta \qquad L^4_{\underset{\vee}{32}} = -\frac{\beta B_4+uA_4}{2\sigma}\sin\theta$$

$$L^1_{\underset{\vee}{14}} = \frac{\sigma}{2w}\left\{\log\left(1-\frac{w^2}{\alpha\sigma}\right)\right\}' \qquad L^4_{\underset{\vee}{14}} = -\frac{\alpha}{2w}\overset{\cdot}{\Big\{}\log\left(1-\frac{w^2}{\alpha\sigma}\right)\Big\} \tag{159}$$

$$L^2_{\underset{\vee}{31}} = L^3_{\underset{\vee}{12}}\sin^2\theta = -\frac{B_1}{2}\sin\theta \qquad L^2_{\underset{\vee}{34}} = -L^3_{\underset{\vee}{24}}\sin^2\theta = -\frac{B_4}{2}\sin\theta$$

$$L^2_{\underset{\vee}{24}} = L^3_{\underset{\vee}{34}} = -\frac{w}{2\alpha}A_1 \qquad L^2_{\underset{\vee}{12}} = -L^3_{\underset{\vee}{31}} = \frac{w}{2\sigma}A_4$$

where

$$A_1 = \frac{uu'+\beta\beta'}{u^2+\beta^2} \qquad B_1 = \frac{u\beta'-\beta u'}{u^2+\beta^2}$$

$$A_4 = \frac{\dot{u}u+\dot{\beta}\beta}{u^2+\beta^2} \qquad B_4 = \frac{u\dot{\beta}-\beta\dot{u}}{u^2+\beta^2} \tag{160}$$

with

$$\{f(r,t)\}' \equiv \frac{\partial f}{\partial r}, \qquad \{f(\dot{r},t)\} \equiv \frac{\partial f}{\partial t}$$

Consider now the first of the field-equations ($\bar{\text{A}}$)

$$\mathfrak{g}^{\alpha\beta}_{\vee\;,\alpha} = 0 \tag{105}$$

From (157) this reduces to

$$(\sqrt{(-g)}\,g^{\underset{\vee}{14}})' = (\sqrt{(-\dot{g})}\,g^{\underset{\vee}{14}}) = 0 \tag{161}$$

with

$$g^{\underset{\vee}{14}} = \frac{-w}{\alpha\sigma-w^2}$$

$$\sqrt{-g} = \sqrt{[(\alpha\sigma-w^2)(\beta^2+u^2)\sin^2\theta]}$$

(161) implies therefore

$$\frac{\beta^2+u^2}{\alpha\sigma-w^2}w^2 = k^2, \qquad k = \text{constant}$$

or

$$\frac{w^2}{\alpha\sigma} = \frac{k^2}{\beta^2+u^2+k^2} \tag{162}$$

The special case $k = 0$ is satisfied by $w = 0$. One now computes $R_{\mu\nu}(L)$ from (158)–(160) and it is seen that the remaining field equations ($\bar{\text{A}}$)

$$R_{\underline{\mu\nu}}(L) = 0$$

$$R_{\underset{\vee}{\mu\nu},\lambda}(L)+R_{\underset{\vee}{\nu\lambda},\mu}(L)+R_{\underset{\vee}{\lambda\mu},\nu}(L) = 0$$

become in the spherically symmetric case

$$\left.\begin{aligned} R_{11}(L) &= 0 \\ R_{22}(L) &= 0 \\ R_{44}(L) &= 0 \\ R_{\underline{14}}(L) &= 0 \end{aligned}\right\} \tag{163}$$

and $\quad R_{\underset{\vee}{23},1}(L) = R_{\underset{\vee}{23},4}(L) = 0 \quad$ or $\quad R_{\underset{\vee}{23}}(L) = \text{constant} \quad$ (164)

The equations $R_{\underset{\vee}{14},1}(L) = R_{\underset{\vee}{14},2}(L) = 0$ are automatically satisfied by $R_{\underset{\vee}{14}}(L)$. Papapetrou† has solved these equations under the assumption $u = 0$, $w \neq 0$. One of the functions α, β, σ, w remains arbitrary, and, therefore, we can set

$$\beta(r, t) = \beta(r) = r^2 \tag{165}$$

We have then

$$A_1 = \frac{\beta'}{\beta} = \frac{2}{r}, \qquad B_1 = A_4 = B_4 = 0$$

$$\frac{w^2}{\alpha\sigma} = \frac{k^2}{k^2+r^4} \tag{166}$$

$$R_{\underset{\vee}{23}}(L) \equiv 0 \quad \text{(identically)} \tag{167}$$

The equation $R_{\underline{14}}(L) = 0$ implies $\dot{\alpha} = 0$ and the remaining field equations (163) become

$$R_{11}(L) = \frac{1}{2}\left(\frac{\sigma'}{\sigma}\right)'+\frac{\sigma'}{4\sigma}\left(\frac{\sigma'}{\sigma}-\frac{\alpha'}{\alpha}\right)-\frac{1}{r}\frac{\alpha'}{\alpha}+2\left(\frac{w^2}{r\alpha\sigma}\right)'$$
$$+\frac{2w^2}{r\alpha\sigma}\left(\frac{\sigma'}{\sigma}-\frac{\alpha'}{2\alpha}+\frac{2w^2}{r\alpha\sigma}\right) = 0 \tag{168}$$

† See footnote, p. 99.

$$R_{22}(L) = \frac{R_{33}}{\sin^2\theta} \equiv \frac{r}{2\alpha}\left(\frac{\sigma'}{\sigma}-\frac{\alpha'}{\alpha}\right)+\frac{1-\alpha}{\alpha}+\frac{2w^2}{\alpha^2\sigma} = 0 \tag{169}$$

$$R_{44}(L) \equiv -\frac{1}{2}\left(\frac{\sigma'}{\alpha}\right)'+\frac{\sigma'}{4\alpha}\left(\frac{\sigma'}{\sigma}-\frac{\alpha'}{\alpha}-\frac{4}{r}\right)-4\left(\frac{w^2}{r\alpha^2}\right)' + \frac{w^2}{r\alpha^2}\left(\frac{3\sigma'}{\sigma}-\frac{2\alpha'}{\alpha}-\frac{14}{r}+\frac{8w^2}{r\alpha\sigma}\right) = 0 \tag{170}$$

These equations do not contain any more the time derivatives of the unknown functions. Hence they are identical with the static case. Forming $R_{11}(L)+(\alpha/\sigma)R_{44}(L)$ we deduce

$$\alpha\sigma = 1+\frac{k^2}{r^4} \tag{171}$$

and therefore, from (166)

$$w^2 = \frac{k^2}{r^4} \tag{172}$$

Equations (168) and (169) reduce to a single relation because

$$\{R_{22}(L)\}' = \frac{2r}{\alpha}R_{11}(L)$$

And (169) becomes

$$1+\frac{r}{2\alpha}(\log\alpha\sigma)'-\frac{r}{\alpha}(\log\sigma)'-\frac{1}{\alpha}-\frac{r}{2\alpha}\left\{\log\left(\frac{r^4}{k^2+r^4}\right)\right\}' = 0 \tag{173}$$

which reduces to a differential equation for σ or α in view of (171). Papapetrou's solution is then (for the case $u = 0$)

$$\beta = r^2 \qquad w = \frac{k}{r^2}$$
$$\alpha = \frac{1}{1-\dfrac{2m}{r}} \qquad \sigma = \left(1+\frac{k^2}{r^4}\right)\left(1-\frac{2m}{r}\right) \tag{174}$$

satisfying the boundary conditions (as $r \to \infty$)

$$\alpha \to 1, \qquad w \to 0, \qquad \beta \to \infty, \qquad \sigma \to 1 \tag{175}$$

The solution contains two constants, m and k, the latter presumably representing the source of the electromagnetic field. It should be noted that the symmetric part of the solution (174) does not coincide with the corresponding solution (156) of the Einstein-Maxwell's equations, even at large distances.

The case $w = 0, u \neq 0$ has been solved by Wyman† in the static case. The field-equations are then

$$A' + \frac{1}{2}(A^2 + B^2) - \frac{1}{2}A\left(\frac{\alpha'}{\alpha} + \frac{\sigma'}{\sigma}\right) = 0 \tag{176}$$

$$\sigma'' - \frac{1}{2}\sigma'\left(\frac{\alpha'}{\alpha} + \frac{\sigma'}{\sigma}\right) + A\sigma' = 0 \tag{177}$$

$$\beta'' - u'B - \frac{1}{2}\beta'\left(\frac{\alpha'}{\alpha} - \frac{\sigma'}{\sigma}\right) + \frac{2\alpha(2\beta uc - \beta^2 + u^2)}{u^2 + \beta^2} = 0 \tag{178}$$

$$u'' + \beta' B - \frac{1}{2}u'\left(\frac{\alpha'}{\alpha} - \frac{\sigma'}{\sigma}\right) - \frac{2\alpha(2\beta u + c\beta^2 - cu^2)}{u^2 + \beta^2} = 0 \tag{179}$$

where c is a constant of integration and $A \equiv A_1$, $B \equiv B_1$. Again, one of the functions remains arbitrary. If we let σ be arbitrary, then the other functions are given by:

For $m \neq 0$:

$$u + i\beta = 4m^2 h / [\{e^a \sigma^{\sqrt{(h)}/2} + e^{-a}\sigma^{-\sqrt{(h)}/2}\}^2 \sigma(c+i)]$$

$$\alpha = (\sigma')^2 (u^2 + \beta^2)/4m^2\sigma \tag{180}$$

with $h = 1 + ih_1, \quad i = \sqrt{-1}$

Here m, h_1 are real constants; a is an arbitrary complex constant of integration. A particular solution of (180), by taking

$$h_1 = 0, \quad e^{2a} = -1, \quad \sigma = 1 - \frac{2m}{r}$$

is due to Papapetrou‡:

$$\beta = r^2, \quad u = -cr^2 \tag{181}$$

For $m = 0$ there are two possibilities:

(i)
$$u + i\beta = h \operatorname{sech}^2 (h^{1/2}x + a)/(c+i) \tag{182}$$

$$\sigma = 1, \quad \alpha = (u^2 + \beta^2)(x')^2$$

(ii)
$$u + i\beta = (i - c)/[(c^2+1)(x+a)^2] \tag{183}$$

$$\sigma = 1, \quad \alpha = (u^2 + \beta^2)(x')^2$$

where in each case x is an arbitrary function of r. As for boundary conditions it is natural to assume that $g_{\mu\nu} \to \eta_{\mu\nu}$ as $r \to \infty$. Depending on the coordinate system used this gives us two types of boundary conditions (as $r \to \infty$):

$$\alpha \to 1, \quad \beta \to r^2, \quad \sigma \to 1, \quad u \to 0 \tag{184}$$

† Wyman, M. (1950) *Can. J. Math.* **2**, 427.
‡ See footnote, page 99.

which may be termed "strong" conditions, if we use polar co-ordinates, or

$$\alpha \to 1, \qquad \beta \to r^2, \qquad \sigma \to 1, \qquad u/r^2 \to 0 \tag{185}$$

which is weaker than (184), if we use Cartesian co-ordinates.

If we apply the strong "boundary" condition (184) to (180) we find that $u \equiv 0$ and the solution degenerates into the Schwarzschild solution containing only the constant m. On the other hand the "weak" boundary condition leads to a solution containing two arbitrary constants m and h_1 which can be interpreted as mass and charge.

The general case $uw \neq 0$ has been studied by Vanstone† and the solution is very similar to Wyman's.

$$\begin{aligned} u+i\beta &= \frac{\lambda c_1}{4}\frac{(i-c)}{(1+c^2)y}\sinh^{-2}\left[\frac{\sqrt{c_1}}{2}(\log y - a)\right] \\ \alpha &= \frac{(u^2+\beta^2)(y')^2}{\lambda y}, \qquad \sigma = \frac{k^2+u^2+\beta^2}{u^2+\beta^2}, \qquad w = \frac{ky'}{\sqrt{\lambda}} \end{aligned} \tag{186}$$

where y is an arbitrary function of r, a is a complex constant, $c_1 = 1+ic_0$, and c_0, c, $\lambda \neq 0$, k are real constants. In this case also the "strong" boundary condition (184) implies $u \equiv 0$. These considerations therefore raise the question whether the symmetric part of the non-symmetric field, that is, $g_{\underline{\mu\nu}}$ represents the "real" metric of the physical space-time, which we shall denote by $a_{\mu\nu}(=a_{\nu\mu})$. Papapetrou assumed $a_{\mu\nu} = g_{\underline{\mu\nu}}$ in connection with his solution and found that it is incompatible with general relativity. In principle, $a_{\mu\nu}$ should be completely determined by $g_{\mu\nu}$ and $\Gamma^{\mu}_{\alpha\beta}$ that is,

$$a_{\mu\nu} = a_{\mu\nu}(g_{\alpha\beta}, \Gamma^{\lambda}_{\nu\tau}) \tag{187}$$

and also, since the generalized field equations reduce to those of general relativity when $g_{\underset{\vee}{\mu\nu}} = 0$, it is natural to demand

$$a_{\mu\nu}(g_{\underline{\alpha\beta}}, \Gamma^{\lambda}_{\underline{\nu\tau}}) = g_{\underline{\alpha\beta}} \tag{188}$$

This still leaves a considerable arbitrariness in the choice of $a_{\mu\nu}$. For example, Wyman‡ has shown that if we construct the following symmetric tensor

$$a_{\mu\nu} = g_{\underline{\mu\nu}} + q_\mu q_\nu \tag{189}$$

where

$$\begin{aligned} q_\mu &= (g_{\mu\lambda} g^{\underline{\lambda\nu}} u_\nu)/(1+\tfrac{1}{2} g_{\underset{\vee}{\alpha\beta}} g^{\underset{\vee}{\alpha\beta}})^{1/2} \\ u_\nu &= h_\nu/(g^{\underline{\rho\sigma}} h_\rho h_\sigma)^{1/2} \\ h_\nu &= g_{\underline{\beta\lambda}} g^{\underset{\vee}{\alpha\beta}} L^{\lambda}_{\underset{\vee}{\alpha\nu}} \end{aligned}$$

† Vanstone, J. R. (1962) *Can. J. Math.* **14**, 568.
‡ See footnote, p. 103.

from the $g_{\mu\nu}$ of Papapetrou, that is, (174), we find that

$$a_{\mu\nu} = \underline{g}_{\mu\nu} \quad \text{for} \quad \mu, \nu \neq 4$$
$$a_{44} = g_{44} + g_{41}^{2} g^{11}/(1+g_{14}{}^{2})$$

or

$$a_{11} = -\left(1-\frac{2m}{r}\right)^{-1}, \quad a_{22} = -r^2, \quad a_{33} = -r^2 \sin^2\theta, \quad a_{44} = 1-\frac{2m}{r}$$

that is, we recover the Schwarzschild solution (155). Thus, if the above $a_{\mu\nu}$ is chosen as the "real" metric Papapetrou's solution of the unified field-equations then really corresponds to a pure gravitational field even though the second constant k appears in the solution. This constant does not appear in the above choice of $a_{\mu\nu}$. Vanstone† has shown that a similar construction is also possible in the general case: $uw \neq 0$. Wyman's construction, in spite of its artificial character, shows not only that a straight-forward identification of $\underline{g}_{\mu\nu}$ with the "real" metric is perhaps too naïve but also that it remains an open problem.

A related problem is that of interpretation of $g_{\mu\nu}$. A comparison of equation (105) with Maxwell's equations suggests that $\underset{\vee}{\mathfrak{h}}^{\beta} \equiv \underset{\vee}{\mathfrak{g}}^{\alpha\beta}{}_{,\alpha}$ be interpreted as the magnetic current density. The electric current density should then be given by

$$\mathfrak{j}^{\mu} = \tfrac{1}{6}\varepsilon^{\alpha\beta\delta\mu}(\underset{\vee}{g}_{\alpha\beta,\delta} + \underset{\vee}{g}_{\beta\delta,\alpha} + \underset{\vee}{g}_{\delta\alpha,\beta}) \tag{190}$$

where $\varepsilon^{\alpha\beta\delta\mu}$ is the Levi-Civita antisymmetric tensor density. Consider now the static spherically symmetric case where the only non-zero components of $\underset{\vee}{g}_{\alpha\beta}$ are $\underset{\vee}{g}_{23}$ and $\underset{\vee}{g}_{14}$ (cf. (157)). The contribution to $\mathfrak{j}^{4}$ then comes from $\underset{\vee}{g}_{23}$ only, and, therefore, one is led to identify, in general, $\underset{\vee}{g}_{ik}(i, k = 1, 2, 3)$ with the electric field **E** and $\underset{\vee}{g}_{i4}$ with the magnetic field **H** respectively, in contrast to the conventional identification (see Part 1, equation (11)) in Maxwell's theory. This by itself is of course not a contradiction. But, if we want to impose the "strong" boundary conditions on spherically symmetric solutions we have seen that u must vanish; so that the only surviving component of $\underset{\vee}{g}_{\alpha\beta}$ is $\underset{\vee}{g}_{14}$, which would imply the existence of isolated magnetic poles. We note, however, that there are other antisymmetric tensors in the theory, for example, $\underset{\vee}{R}_{\mu\nu}$ and $R^{\underset{\vee}{\mu\nu}}$ where $R^{\mu\nu}$ is defined by‡

$$R^{\mu\lambda}R_{\nu\lambda} = R^{\lambda\mu}R_{\lambda\nu} = \delta^{\mu}_{\nu} \tag{191}$$

There are therefore four antisymmetric tensors $g^{\underset{\vee}{\mu\nu}}$, $\underset{\vee}{g}_{\mu\nu}$, $\underset{\vee}{R}_{\mu\nu}$, $R^{\underset{\vee}{\mu\nu}}$ which could

† See footnote, p. 104.
‡ Henceforth we set $R_{\mu\nu}(L) \equiv R_{\mu\nu}$.

be in some way associated with the electromagnetic field, and two symmetric tensors $g_{\mu\nu}$, $g^{\underline{\mu\nu}}$ which could be associated with the metric of space-time. From $\overline{g^{\underline{\mu\nu}}}$ we can define a covariant symmetric tensor $h_{\mu\nu}$ by

$$g^{\underline{\mu\lambda}} h_{\nu\lambda} = \delta^{\mu}_{\nu} \tag{192}$$

Thus the metric could also be identified with $h_{\mu\nu}$ instead of $g_{\mu\nu}$. Following Schrödinger† one could, for example, start from an action function

$$\overline{\mathscr{W}} \equiv \overline{\mathscr{W}}(h_{\mu\nu}, R_{\underset{\vee}{\mu\nu}}) = 2\alpha\{\sqrt{-|(h_{\mu\nu} + R_{\underset{\vee}{\mu\nu}})|} - \sqrt{(-|h_{\mu\nu}|)}\} \tag{193}$$

and define the conjugate quantities by

$$R_{\underline{\mu\nu}} = \frac{\partial \overline{\mathscr{W}}}{\partial \mathfrak{g}^{\underline{\mu\nu}}}, \qquad \mathfrak{g}^{\underset{\vee}{\mu\nu}} = \frac{\partial \overline{\mathscr{W}}}{\partial R_{\underset{\vee}{\mu\nu}}} \tag{194}$$

Here α is a constant. (193) and (194) imply

$$\mathfrak{g}^{\underset{\vee}{\mu\nu}} = \frac{\alpha\sqrt{-h}}{A}(g^{\underline{\mu\rho}} g^{\underline{\nu\sigma}} R_{\underset{\vee}{\rho\sigma}} - I_2 R^{*\underset{\vee}{\mu\nu}}) \tag{195}$$

$$R_{\underline{\mu\nu}} = \alpha/A[g^{\underline{\rho\sigma}} R_{\underset{\vee}{\mu\rho}} R_{\underset{\vee}{\nu\sigma}} - h_{\mu\nu}(A-1)] \tag{196}$$

where $$R^{*\underset{\vee}{\mu\nu}} = \frac{1}{2\sqrt{-h}} \varepsilon^{\mu\nu\rho\sigma} R_{\underset{\vee}{\rho\sigma}}, \qquad I_2 = \left(\frac{1}{4}\right) R^{*\underset{\vee}{\mu\nu}} R_{\underset{\vee}{\mu\nu}}$$

$$A = [1 + \tfrac{1}{2} g^{\underline{\mu\rho}} g^{\underline{\nu\sigma}} R_{\underset{\vee}{\mu\nu}} R_{\underset{\vee}{\rho\sigma}} - I_2{}^2]^{1/2}$$

$$h = |h_{\mu\nu}|$$

(195) and (196) are to be regarded as field-equations in this version of the theory. A satisfactory feature of the above field-equations is that (196) enables us to express $R_{\underline{\mu\nu}} - \frac{1}{2} g^{\underline{\rho\sigma}} h_{\mu\nu} R_{\underline{\rho\sigma}}$ in terms of an energy tensor $E_{\mu\nu}$ of the Maxwellian type (see equation $\overline{(1)}$), where the electromagnetic field is represented by $R_{\underset{\vee}{\mu\nu}}$. On the other hand, in (195) the relationship between the contravariant tensor density $\mathfrak{g}^{\underset{\vee}{\mu\nu}}$ and $R_{\underset{\vee}{\mu\nu}}$ is similar to the corresponding relationship between the field and induction in Born-Infeld theory. In fact, in the static spherically symmetric case a solution of (105) and (195)–(196) of the form

$$h_{11} = -e^{\lambda(r)}, \qquad h_{22} = -r^2, \qquad h_{33} = -r^2 \sin^2\theta, \qquad h_{44} = e^{\nu(r)}$$

$$R_{\underset{\vee}{41}} = -R_{\underset{\vee}{14}} = \psi(r)$$

exists, where

$$e^{\nu(r)} = e^{-\lambda(r)} = 1 - \frac{f^2}{r}\int_0^r (\sqrt{(\varepsilon^2 + r^4)} - r^2)\, dr$$

$$\psi(r) = b\varepsilon/c\sqrt{(\varepsilon^2 + r^4)} \tag{197}$$

and f, b, ε, c are constants.

† Schrödinger, E. (1943) *Proc. R. Ir. Acad.* **49**A. 43, 225. Schrödinger's formalism was originally in a *symmetric pure* affine framework.

That is, the field remains finite at the origin $r = 0$, as in the theory of Born-Infeld. Moreover, for $r \gg \sqrt{\varepsilon}$ the solution has the Schwarzschild form:

$$e^{\nu} = e^{-\lambda} = 1 - \frac{2k}{c^2}\frac{m_0}{r}$$

Only *in this (or similar) version of the theory can one say that the field-particle duality is absent in the non-symmetric field theory.* On the other hand, if particles are represented as point-singularities of the field, one should expect that the method of Einstein-Infeld-Hoffmann (see p. 19) when applied to the field-equations of the non-symmetric theory should lead to the Lorentz equations of motion of a charged particle in an electromagnetic field. It turns out to be not the case.†

The Einstein-Infeld-Hoffmann method employs a quasi-static approximation procedure in which the time derivatives are of an order smaller than the space derivatives. Expanded in power series of ϵ, we have then ($i, k = 1, 2, 3$):

$$\begin{aligned}
g_{\underline{44}} &= 1 + \epsilon^2 \underset{2}{g}_{\underline{44}} + \epsilon^4 \underset{4}{g}_{\underline{44}} + \cdots \\
g_{\underline{4k}} &= \epsilon^3 \underset{3}{g}_{\underline{4k}} + \epsilon^5 \underset{5}{g}_{\underline{4k}} + \cdots \\
g_{\underline{ik}} &= -\delta_{ik} + \epsilon^2 \underset{2}{g}_{\underline{ik}} + \epsilon^4 \underset{4}{g}_{\underline{ik}} + \cdots \qquad (198) \\
g_{\underset{\vee}{i4}} &= \epsilon^3 \underset{3}{g}_{\underset{\vee}{i4}} + \epsilon^5 \underset{5}{g}_{\underset{\vee}{i4}} + \cdots \\
g_{\underset{\vee}{ik}} &= \epsilon^2 \underset{2}{g}_{\underset{\vee}{ik}} + \epsilon^4 \underset{4}{g}_{\underset{\vee}{ik}} + \cdots
\end{aligned}$$

From (127), (128) and above we obtain for the connection components in the various orders:

$$\begin{aligned}
&\underset{3}{u}^4_{44} = 0 \qquad && \underset{2}{u}^p_{44} = \underset{2}{u}^4_{p4} = \underset{2}{L}^4_{\underset{\vee}{p4}} = 0 \\
&\underset{3}{L}^q_{\underset{\vee}{4p}} = -\underset{3}{g}_{\underset{\vee}{4p},q} + \underset{3}{h}_{4pq} \qquad && \underset{3}{L}^4_{\underset{\vee}{pq}} = \underset{3}{g}_{\underset{\vee}{pq},4} - \underset{3}{h}_{pq4} \\
&\underset{2}{L}^p_{\underset{\vee}{qr}} = -\underset{2}{g}_{\underset{\vee}{qr},p} + \underset{2}{h}_{qrp} \qquad && \underset{4}{L}^4_{\underset{\vee}{4p}} = \underset{4}{u}^4_{4p} = \underset{4}{u}^p_{44} = 0
\end{aligned} \qquad (199)$$

$$\underset{4}{u}^r_{pq} = -\underset{2}{g}_{\underset{\vee}{pm}}(-\underset{2}{g}_{\underset{\vee}{qr},m} + \underset{2}{h}_{qrm}) - \underset{2}{g}_{\underset{\vee}{mq}}(-\underset{2}{g}_{\underset{\vee}{rp},m} + \underset{2}{h}_{rpm})$$

$$\underset{4}{L}^r_{\underset{\vee}{pq}} = -\underset{4}{g}_{\underset{\vee}{pq},r} + \underset{4}{h}_{pqr} + \underset{2}{g}^{\underline{rs}}(\underset{2}{g}_{\underset{\vee}{pq},s} - \underset{2}{h}_{pqs}) + \underset{2}{g}_{\underset{\vee}{ps}}\underset{2}{\left\{ {s \atop rq} \right\}} - \underset{2}{g}_{\underset{\vee}{qs}}\underset{2}{\left\{ {s \atop pr} \right\}}$$

† Callaway, J. (1953) *Phys. Rev.* **92**, 1567.

where
$$h_{\alpha\beta\gamma} \equiv g_{\underset{\vee}{\alpha\beta},\gamma}+g_{\underset{\vee}{\beta\gamma},\alpha}+g_{\underset{\vee}{\gamma\alpha},\beta}$$
$$\left\{ {s \atop rq} \right\} \equiv \left\{ {s \atop rq} \right\}_{g_{\underline{\mu\nu}}}$$

We can then calculate the contracted curvature tensor in various orders. Substituting in the field-equations we obtain the following equations for the second- and third-order antisymmetric components in the weak-field approximation

$$\begin{aligned} \nabla^2(\underset{2}{g}_{\underset{\vee}{ik},l}+\underset{2}{g}_{\underset{\vee}{kl},i}+\underset{2}{g}_{\underset{\vee}{li},k}) &= 0 \\ \nabla^2(\underset{3}{g}_{\underset{\vee}{4k},l}+\underset{3}{g}_{\underset{\vee}{kl},4}+\underset{3}{g}_{\underset{\vee}{l4},k}) &= 0 \end{aligned} \tag{200}$$

And (105) leads us to

$$\begin{aligned} \operatorname{div} \underset{2}{g}_{\underset{\vee}{ik}} &= \underset{2}{g}_{\underset{\vee}{ik},k} = 0 \\ \operatorname{div} \underset{3}{g}_{\underset{\vee}{4k}} &= \underset{3}{g}_{\underset{\vee}{4k},k} = 0 \end{aligned} \tag{201}$$

We now look for particle type solutions of the above equations which will reduce, at distances large compared to the gravitational radius of the particles, to the ordinary electric and magnetic fields of slowly moving charges. Thus we introduce the potential (for a system of N-particles)

$$\underset{2}{\varphi} = \sum_{i=1}^{N} \underset{2}{\varphi}(i)$$

Where $\varphi(i) = e(i)/r(i)$, $r(i)$ being the distance of the ith particle and $e(i)$ a constant proportional to the charge. If we assume that $g_{\underset{\vee}{ik}}$ represents the electric field (see p. 105) and $g_{\underset{\vee}{4k}}$ the magnetic field we must set

$$\underset{2}{g}_{\underset{\vee}{ik}} = \varepsilon_{ikl}\underset{2}{\varphi}_{,l} \tag{202}$$

where ε_{ikl} is Levi-Civita symbol in 3-space. This satisfies the field-equations to the second order. To satisfy the field-equations in the third order we denote the motion of the ith singularity by three functions of time $\xi^k(i, t)$ with velocities of order ϵ, $\dot{\xi}^k(i, t)$, and accelerations of order ϵ^2, $\ddot{\xi}^k(i, t)$. Then

$$\underset{3}{g}_{\underset{\vee}{4k}} = -\varepsilon_{4kij}\left(\sum_{l=1}^{N} \underset{2}{\varphi}(l)\dot{\xi}_i(l, t)\right)_{,j} \tag{203}$$

satisfies the field equations in the third order.

In general relativity the equations of motion of the kth singularity in a gravitational field are obtained by first surrounding the singularity by a closed surface and considering the surface integral of certain quantities Λ_{ik} (which are related to the Ricci tensor). The surface integrals are independent of the surface and depend only on the co-ordinates of the singularities thereby furnishing the equations of motion. The equations which one must consider here are

$$\underline{R_{\mu\nu}} = 0, \qquad R_{\underset{\vee}{\mu\nu},\lambda}+R_{\underset{\vee}{\nu\lambda},\mu}+R_{\underset{\vee}{\lambda\mu},\nu} = 0$$

The second of the equations show that $R_{\underset{\vee}{\mu\nu}}$, being the curl of a vector cannot contribute anything to the equations of motion. The corresponding quantity Λ_{mn} in the unified field theory is given by

$$-\underset{4}{\Lambda}_{mn} = \{\underset{2}{g}_{\underset{\vee}{ms}}\underset{2}{g}_{\underset{\vee}{np},s}+\delta_{pm}\underset{2}{\varphi}_{,r,n}\underset{2}{\varphi}_{,r}-\delta_{mn}\underset{2}{\varphi}_{,p,r}\underset{2}{\varphi}_{,r}\}_{,p} \tag{204}$$

with the property $\underset{4}{\Lambda}_{mn,n} = 0$, so that the integral over the surface Ω^k surrounding the kth singularity

$$\underset{4}{C}_m(k) \equiv \int_{\overset{k}{\Omega}} \underset{4}{\Lambda}_{mn} n_n \, d\Omega$$

does not depend on the surface. The equations of motion should be then

$$\underset{4}{C}_m(k) = 0 \tag{205}$$

However, from (204) it can be seen that $\underset{4}{C}_m(k)$ vanishes identically. Therefore, we cannot obtain the Lorentz equation of motion from the field-equations ($\bar{\text{A}}$) if we represent charged particles as point-singularities. A similar analysis† applied to the system (B) shows also a negative result. This would seem to strengthen the non-dualistic viewpoint that particles should be represented by *non-singular* solutions of the field equations.‡

The non-symmetric theory in spite of its initial promise cannot be considered to have solved the problem of unification or field-particle duality in a complete and satisfactory manner.

2. *Wheeler-Misner's geometrodynamics*

In a classical field theory, it would seem, there are only two ways to represent matter (particles)—either as a singularity of the field or (as in Mie-Born-Infeld type of theories) as a distribution of field energy diffused

† Infeld, L. (1950) *Acta Phys. Pol.* **10**, 284.

‡ Bonnor, W. B. (1954) in *Proc. R. Soc.* **226**A, 366, shows that a modified action function

$$\mathscr{W}^* = \sqrt{(-g)}\, g^{\mu\nu} R_{\mu\nu} + p^2 \sqrt{(-g)}\, g^{\mu\nu} g_{\mu\nu}$$

where p is a constant can lead to Lorentz equations of motion according to the Einstein-Infeld-Hoffmann method.

throughout the space. There exists, however, another possibility as shown by Wheeler and Misner.† This is to bring in the *topology* of the space-time manifold to play a decisive role to explain matter in a geometrized field theory. Wheeler and Misner consider the Einstein-Maxwell's equations in *empty space* (that is, equations (1) without s_μ, $T_{\mu\lambda}$) and show, first of all, that they can be regarded as equations characterizing an "already" unified and geometrized theory of gravitation and electromagnetism—a result first obtained by Rainich‡ in 1925. Next, they note that the field-equations have a purely local character and therefore indicate nothing about the topology of the manifold in the large. It is possible, therefore, at least qualitatively, to represent matter as "wormlike" regions of a multiply-connected space-time which is otherwise endowed with a *everywhere-regular Riemannian* metric.

Recall that the *free-field* Einstein-Maxwell's equations are (see equation (1))

$$R_{\mu\lambda}-\tfrac{1}{2}g_{\mu\lambda}R = -\kappa E_{\mu\lambda} = 2(f_{\mu\nu}f^{\nu}_{\ \lambda}-\tfrac{1}{2}g_{\mu\lambda}f_{\alpha\beta}f^{\alpha\beta}) \tag{206}$$

$$f^{\mu\lambda}_{\ \ ;\lambda} \equiv (-g)^{-1/2}(f^{\mu\lambda}\sqrt{-g})_{,\lambda} = 0 \tag{207}$$

$$f_{\mu\lambda,\nu}+f_{\lambda\nu,\mu}+f_{\nu\mu,\lambda} = 0 \tag{208}$$

where on the right-hand side of (206) we have absorbed the constant κ in the field $f_{\mu\lambda}$. Here

$$f^{\alpha\beta} = g^{\alpha\mu}g^{\beta\nu}f_{\mu\nu} \tag{209}$$

Equation (208) is equivalent to

$$*f^{\beta\alpha}_{\ \ ;\alpha} \equiv (1/\sqrt{-g})(*f^{\beta\alpha}\sqrt{-g})_{,\alpha} = 0 \tag{210}$$

where

$$*f^{\beta\alpha} = \frac{1}{2\sqrt{-g}}\,\varepsilon^{\beta\alpha\mu\nu}f_{\mu\nu} \tag{211}$$

is the dual electromagnetic field.

Consider first the relation between the electromagnetic field **f** and the electromagnetic energy-momentum tensor $E_{\mu\lambda}$. This relation is purely algebraic; so that we may concentrate our attention on a single space-time point and use a co-ordinate system in which $g_{\mu\lambda}$ assumes the Minkowski values at that point. One can form two principal invariants out of $f_{\mu\lambda}$ and $*f_{\mu\lambda}$:

$$\mathbf{f}^2 = \tfrac{1}{2}f_{\mu\lambda}f^{\mu\lambda} \quad (=\mathbf{H}^2-\mathbf{E}^2 \text{ in a Minkowski frame}) \tag{212}$$

$$\mathbf{f}\times\mathbf{f} = \tfrac{1}{2}f_{\mu\lambda}{*f^{\mu\lambda}} \quad (=2(\mathbf{E}.\mathbf{H}) \text{ in a Minkowski frame}) \tag{213}$$

† See footnote in Introduction.
‡ Rainich, G. Y. (1925) *Trans. Am. Math. Soc.* **27**, 106.

The electromagnetic energy momentum tensor can then be expressed more symmetrically as follows

$$\tau(\mathbf{f}) \equiv 2(f_{\mu\alpha}f^{\alpha}_{\ \nu}-\tfrac{1}{4}g_{\mu\nu}f_{\alpha\beta}f^{\alpha\beta}) = f_{\mu\alpha}f^{\alpha}_{\ \nu}+{}^*f_{\mu\alpha}{}^*f^{\alpha}_{\ \nu} \tag{214}$$

because of the identity†

$$f_{\mu\alpha}f^{\nu\alpha}-{}^*f_{\mu\alpha}{}^*f^{\nu\alpha} = \tfrac{1}{2}\delta^{\nu}_{\mu}f_{\alpha\beta}f^{\alpha\beta} \tag{215}$$

Wheeler and Misner call (214) the "Maxwell square of **f**". $\tau(\mathbf{f})$ has also the following symmetry. Consider an angle α and define the operation $e^{*\alpha}$ by

$$e^{*\alpha}\mathbf{f} = \mathbf{f}\cos\alpha+{}^*\mathbf{f}\sin\alpha \tag{216}$$

Equation (216) can be considered as a *duality rotation*. It has the additivity property

$$e^{*\alpha}e^{*\beta} = e^{*\beta}e^{*\alpha} = e^{*(\alpha+\beta)} \tag{217}$$

The dual of the duality rotation gives us

$$^*(e^{*\alpha}\mathbf{f}) = -\mathbf{f}\sin\alpha+{}^*\mathbf{f}\cos\alpha \tag{218}$$

because

$$^*(^*\mathbf{f}) = -\mathbf{f} \quad \text{or} \quad {}^*({}^*f_{\alpha\beta}) = -f_{\alpha\beta} \tag{219}$$

$\tau(\mathbf{f})$ has the following property

$$\tau(e^{*\alpha}\mathbf{f}) = \tau(\mathbf{f}) \tag{220}$$

that is, the Maxwell square of a duality-rotated field is identical with the Maxwell square of the original field. Let

$$\xi = e^{-*\alpha}\mathbf{f} \tag{221}$$

be a field which has undergone a duality rotation by an angle $-\alpha$. Then

$$\xi^2 = \mathbf{f}^2\cos 2\alpha-\mathbf{f}\times\mathbf{f}\sin 2\alpha$$

$$\xi\times\xi = \mathbf{f}^2\sin 2\alpha+\mathbf{f}\times\mathbf{f}\cos 2\alpha \tag{222}$$

Assume that the invariants $\mathbf{f}^2$ and $\mathbf{f}\times\mathbf{f}$ of the original field do not both vanish. Then, one can choose the angle α such that

$$\xi\times\xi = 0 \tag{223}$$

that is,

$$\tan 2\alpha = -(\mathbf{f}\times\mathbf{f})/\mathbf{f}^2 \tag{224}$$

Then from (222),

$$\xi^2 = \pm[(\mathbf{f}^2)^2+(\mathbf{f}\times\mathbf{f})^2]^{1/2} \tag{225}$$

† $a_{\mu\alpha}b^{\nu\alpha}-{}^*a_{\mu\alpha}{}^*b^{\nu\alpha} = \frac{1}{2}\delta^{\nu}_{\mu}a_{\alpha\beta}b^{\alpha\beta}$ for any two antisymmetric tensors $a_{\alpha\beta}$, $b_{\alpha\beta}$.

By demanding the *negative* sign on the right-hand side of (225) we can determine the angle 2α uniquely (up to a positive or negative additive integral multiple of 2π) from the field $\mathbf{f}$ according to (224).

Equations (221) and (223) define a duality rotation of a given field $\mathbf{f}$ into an *extremal* field $\boldsymbol{\xi}$, which represents, in a Minkowski frame, pure electric field or a Lorentz transformation thereof. Referred to an extremal field $\boldsymbol{\xi}$, the actual field $\mathbf{f}$ then arises out of a duality rotation by an angle α.

$$\mathbf{f} = e^{*\alpha}\boldsymbol{\xi} \quad \text{or} \quad f_{\mu\lambda} = e^{*\alpha}\xi_{\mu\lambda} \tag{226}$$

The angle α can be called the "*complexion*" of the field $\mathbf{f}$. When the field $\mathbf{f}$ is a *null* field i.e.,

$$\mathbf{f}^2 = \mathbf{f}\times\mathbf{f} = 0 \tag{227}$$

the equation (224) for α becomes indeterminate. Then the complexion cannot be defined on a purely local basis.

Consider now the energy-momentum tensor for the non-null case. We have, in view of (220)

$$\begin{aligned} \tau_\mu^\lambda(\mathbf{f}) &= \tau_\mu^\lambda(\boldsymbol{\xi}) = 2\xi_{\mu\alpha}\xi^{\lambda\alpha} - \delta_\mu^\lambda(\xi^2) \\ \tau_\lambda^\nu(\mathbf{f}) &= \tau_\lambda^\nu(*\boldsymbol{\xi}) = 2{}^*\xi_{\lambda\alpha}{}^*\xi^{\nu\alpha} + \delta_\lambda^\nu(\xi^2) \end{aligned} \tag{228}$$

Using (215) and (223) one obtains from above

$$\tau_\mu^\lambda\tau_\lambda^\nu \equiv \delta_\mu^\nu(\xi^2)^2 = \delta_\mu^\nu[(\mathbf{f}^2)^2 + (\mathbf{f}\times\mathbf{f})^2] \tag{229}$$

In the null case the right-hand side vanishes. Thus the square of the energy-momentum tensor is in general a multiple of the identity matrix. From (206), in terms of the contracted curvature tensor, we have then

$$R_\mu^\lambda R_\lambda^\nu = \tfrac{1}{4}\delta_\mu^\nu(R_{\alpha\beta}R^{\alpha\beta}) \tag{230}$$

The value of the constant is obtained by comparing the traces of the two sides of the equation. On the other hand τ_μ^λ has the property that its trace vanishes, that is,

$$\tau_\lambda^\lambda = 0 \tag{231}$$

Moreover, the electromagnetic energy density, given by the component τ_{00} must be positive definite. In terms of the contracted curvature tensor, therefore, again from (206), we require

$$R_\lambda^\lambda \equiv R = 0 \tag{232}$$

$$R_{00} \geqslant 0 \tag{233}$$

Equations (230), (232) and (233) are the Rainich conditions which $R_{\mu\lambda}$ must satisfy if it is to be representable as the Maxwell square of an antisymmetrical field tensor.

The Rainich conditions now enable us to solve (206) for $f_{\mu\lambda}$ via the extremal field $\xi_{\mu\lambda}$ in terms of the contracted curvature tensor $R_{\mu\lambda}$. Consider the non-null case first, for which $\xi^2 \neq 0$; in other words (from (206), (225) and (229)), the case:

$$\xi^2 = -\tfrac{1}{2}(R_{\alpha\beta}R^{\alpha\beta})^{1/2} \neq 0 \tag{234}$$

or

$$R_{\alpha\beta}R^{\alpha\beta} \neq 0 \tag{235}$$

Let

$$E_{\tau\sigma}{}^{\mu\nu} \equiv \tfrac{1}{2}(-\delta^{\mu}_{\tau}R^{\nu}_{\sigma}+\delta^{\mu}_{\sigma}R^{\nu}_{\tau}-\delta^{\nu}_{\sigma}R^{\mu}_{\tau}+\delta^{\nu}_{\tau}R^{\mu}_{\sigma}) \tag{236}$$

so that,

$$E_{\tau\alpha}{}^{\alpha\nu} = R^{\nu}_{\tau} \tag{237}$$

in view of (232). From (206) and (228) one obtains

$$E_{\alpha\beta\gamma\delta} = -\xi_{\alpha\beta}\xi_{\gamma\delta}-\xi^{*}{}_{\alpha\beta}{}^{*}\xi_{\gamma\delta} \tag{238}$$

and from (234) and (236)

$$E_{\alpha\beta\gamma\delta}E_{\mu\nu}{}^{\gamma\delta} = (R_{\sigma\rho}R^{\sigma\rho})^{1/2}(-\xi_{\alpha\beta}\xi_{\mu\nu}+{}^{*}\xi_{\alpha\beta}{}^{*}\xi_{\mu\nu}) \tag{239}$$

so that, finally from above

$$\xi_{\mu\nu}\xi_{\sigma\tau} = -\tfrac{1}{2}E_{\mu\nu\sigma\tau}-\tfrac{1}{2}(R_{\alpha\beta}R^{\alpha\beta})^{-1/2}E_{\mu\nu\gamma\delta}E_{\sigma\tau}{}^{\gamma\delta} \tag{240}$$

To find any given component of $\xi_{\mu\nu}$ one sets the indices $(\mu, \nu) = (\sigma, \tau)$ and takes the square root of the right-hand side. (240) itself can then be used to determine the relative sign of the different components $\xi_{\mu\nu}$ and $\xi_{\sigma\tau}$.

Consider now the null case for which

$$R_{\mu\nu}R^{\mu\nu} = 0 \tag{241}$$

The contracted curvature tensor then has the form

$$R_{\mu\nu} = 2k_{\mu}k_{\nu}$$

where

$$k_{\alpha}k^{\alpha} = 0 \tag{242}$$

The Maxwell square root of the contracted curvature tensor is simply (apart from a duality rotation)

$$f_{\mu\nu} = k_{\mu}v_{\nu}-k_{\nu}v_{\mu} \tag{243}$$

where

$$k_{\alpha}v^{\alpha} = 0; \qquad v_{\alpha}v^{\alpha} = 1$$

Having expressed the electromagnetic field as the Maxwell square root of the Ricci tensor, where

$$f_{\mu\nu} = e^{*\alpha}\xi_{\mu\nu} \quad \text{or} \quad \mathbf{f} = e^{*\alpha}\boldsymbol{\xi} \tag{226}$$

$$\boldsymbol{\xi} = \text{extremal square root of } R_{\mu\nu}$$

we next substitute the expression in the Maxwell's equations, thereby obtaining relations involving purely geometric quantities. The two sets of Maxwell's equations are

$$f^{\mu\nu}{}_{;\nu} = 0 \tag{207}$$

$$*f^{\mu\nu}{}_{;\nu} = 0 \tag{210}$$

(226) substituted in (207) and (210) gives us

$$\xi^{\mu\nu}{}_{;\nu} + *\xi^{\mu\nu}\alpha_{,\nu} = 0 \tag{244}$$

$$*\xi^{\mu\nu}{}_{;\nu} - \xi^{\mu\nu}\alpha_{,\nu} = 0 \tag{245}$$

Multiplying (244) by $\xi_{\alpha\mu}$ and (245) by $*\xi_{\alpha\mu}$ we get ($\xi_{\alpha\mu}*\xi^{\mu\nu} = \frac{1}{2}\delta^{\mu}_{\alpha}(\xi \times \xi) = 0$)

$$\xi_{\alpha\mu}\xi^{\mu\nu}{}_{;\nu} = 0 \tag{246}$$

$$*\xi_{\alpha\mu}*\xi^{\mu\nu}{}_{;\nu} = 0 \tag{247}$$

From the identities†

$$\begin{aligned} \xi_{\alpha\mu;\nu}\xi^{\mu\nu} &= *\xi_{\alpha\mu}*\xi^{\mu\nu}{}_{;\nu} - \tfrac{1}{2}(\xi^2)_{;\alpha} \\ *\xi_{\alpha\mu;\nu}*\xi^{\mu\nu} &= \xi_{\alpha\mu}\xi^{\mu\nu}{}_{;\nu} + \tfrac{1}{2}(\xi^2)_{;\alpha} \end{aligned} \tag{248}$$

and (246)–(247) we get, therefore,

$$\tfrac{1}{2}(\xi_{\alpha\mu}\xi^{\mu\nu} + *\xi_{\alpha\mu}*\xi^{\mu\nu})_{;\nu} = 0 \tag{249}$$

which in terms of contracted curvature tensor can be written as (since $R = 0$)

$$(R^{\nu}_{\mu} - \tfrac{1}{2}\delta^{\nu}_{\mu}R)_{;\nu} = 0 \tag{250}$$

These are just the Bianchi identities (Part 1, equation (64)). On the other hand, multiplying (244) by $*\xi_{\beta\mu}$ and (245) by $\xi_{\beta\mu}$ and adding we get (from (215))

$$*\xi_{\beta\mu}\xi^{\mu\nu}{}_{;\nu} + \xi_{\beta\mu}*\xi^{\mu\nu}{}_{;\nu} + \delta^{\mu}_{\beta}\xi^2\alpha_{,\nu} = 0 \tag{251}$$

or

$$\alpha_{\beta} = \alpha_{,\beta} \tag{252}$$

where

$$\alpha_{\beta} \equiv -(*\xi_{\beta\mu}\xi^{\mu\nu}{}_{;\nu} + \xi_{\beta\mu}*\xi^{\mu\nu}{}_{;\nu})/\xi^2 \tag{253}$$

Here, of course, we are considering the non-null case $\xi^2 \neq 0$. The right-hand side of (253) can now be expressed in terms of $R_{\mu\lambda}$ by noting that

$$\begin{aligned} E^{\gamma\delta\beta\tau}{}_{;\tau} &= \tfrac{1}{2}(R^{\delta\beta;\gamma} - R^{\gamma\beta;\delta}) \\ &= (-\xi^{\gamma\delta}\xi^{\beta\tau} - *\xi^{\gamma\delta}*\xi^{\beta\tau})_{;\tau} \end{aligned} \tag{254}$$

† $\frac{1}{2}a^{\alpha\beta}(b_{\mu\alpha;\beta} + b_{\alpha\beta;\mu} + b_{\beta\mu;\alpha}) = *a_{\mu\alpha}*b^{\alpha\beta}{}_{;\beta}$ for any two antisymmetric tensors $a_{\alpha\beta}$, $b_{\alpha\beta}$.

and

$$\tfrac{1}{2}(-g)^{1/2}\varepsilon_{\gamma\delta\mu\nu}E_{\alpha\beta}{}^{\mu\nu} = \tfrac{1}{2}(-g)^{1/2}\varepsilon_{\gamma\delta\mu\nu}(\delta^{\nu}_{\alpha}R^{\mu}_{\beta}-\delta^{\nu}_{\beta}R^{\mu}_{\alpha})$$
$$= -\xi_{\alpha\beta}{}^{*}\xi_{\gamma\delta}+{}^{*}\xi_{\alpha\beta}\xi_{\gamma\delta} \qquad (255)$$

Therefore, multiplying (254) by (255) we get

$$\tfrac{1}{2}(-g)^{1/2}\varepsilon_{\alpha\delta\gamma\mu}R^{\delta\beta;\gamma}R^{\mu}_{\beta} = -2\xi^{2}({}^{*}\xi_{\alpha\beta}\xi^{\beta\tau}{}_{;\tau}+\xi_{\alpha\beta}{}^{*}\xi^{\beta\tau}{}_{;\tau})$$

or

$$\alpha_{\beta} = (-g)^{1/2}\varepsilon_{\beta\alpha\mu\nu}R^{\alpha\gamma;\mu}R^{\nu}_{\gamma}/R_{\sigma\tau}R^{\sigma\tau} \qquad (256)$$

From (252) we see that α_{β} is the gradient of the complexion α; consequently

$$\alpha_{\beta;\mu}-\alpha_{\mu;\beta} = \alpha_{\beta,\mu}-\alpha_{\beta,\mu} = 0 \qquad (257)$$

Conversely, when (257) is satisfied the complexion α is determined, up to a constant α_0, from α_{β} by

$$\alpha = \int_{0}^{x}\alpha_{\beta}\,dx^{\beta}+\alpha_{0} \qquad (258)$$

provided the region of the line integral is simply connected and does not include points where $R_{\alpha\beta}R^{\alpha\beta} = 0$. For a *multiply-connected* region it is necessary to replace (257) by the condition that the line integral of α_{β} around any closed path be an integral multiple of 2π

$$\oint\alpha_{\beta}\,dx^{\beta} = 2\pi n \qquad (259)$$

In summary then, equations (256), (259) or (251) together with the algebraic conditions (230), (232) and (233) translate in purely geometric terms the content of the *free-field* Einstein-Maxwell equations (206)–(208), and, therefore, can be considered to comprise an "already" unified and geometrized theory of gravitational and electromagnetic fields.

One could, at this stage, adopt the usual dualistic viewpoint by admitting singularities in the theory and identify the singularities as charged particles. Wheeler and Misner suggest the radical possibility of geometrizing away even the singularities by introducing multiply-connected space-times.

Imagine then a multiply-connected space-time with "handles" where there is a net flux of lines of force entering one hole ("wormhole") and emerging from the other hole of the handle (see figure). Lines of force do not *end* or *begin* at any singular point of space-time. But, if one disregards the dotted portion of the handle the pattern of lines of force is that of a positive and negative elementary electric charge. The *charge* or net flux surrounding the mouth of a wormhole can have any value. The concentration of electromagnetic energy connected with the trapped lines of force in a wormhole imparts *mass* to the region of space. The Einstein-Maxwell equations also predict the possibility of long-lived concentration of electromagnetic energy even in a simply connected space-time. Such regions, called "geons" would correspond to uncharged particles with mass. Both in the simply- and multiply-connected cases the mass is *classical*, *non-localized*, and *unquantized*. It has nothing to do with the quantized mass of the elementary particles.

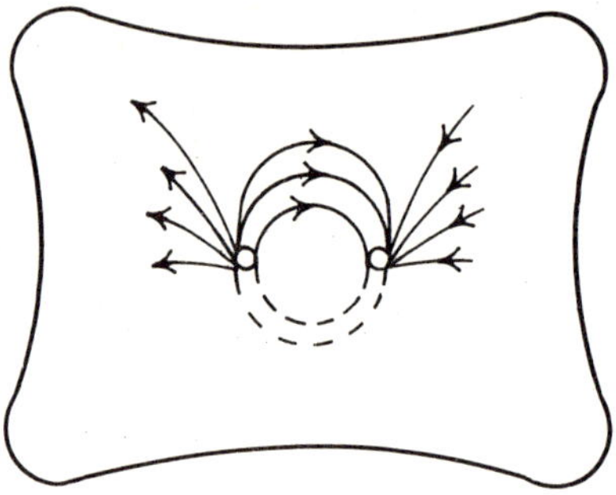

In Wheeler-Misner's geometrodynamics, mass and charge are related to the topological aspects of the geometry of space-time, and, therefore, the equations of motion of charged particles, that is, equations describing temporal change of the topology of the space-time should be, in principle, also derivable from the basic equations of the theory. Such a difficult program has not yet been carried out. At the present stage one has no other recourse but to fall back on the dualistic viewpoint, by idealizing a small handle as a point singularity in a simply-connected continuum and make an appeal to the Einstein-Infeld-Hoffmann technique to derive the equations of motion of such singularities.

Notwithstanding the rather qualitative aspect of Wheeler and Misner's interpretation, it is an example of some of the more imaginative efforts made towards the program of geometrization of physics.

B. Quantum Theory

1. Heisenberg's unified field theory

The elementary particle spectrum. A unified and comprehensive theory of matter should be able to account for the incredible spectrum of elementary particles found in nature, together with their individual and collective properties. At one time it was believed there were altogether only three elementary particles—*proton*, *electron* and the *neutron* and their antiparticles. (To these one must also add the *massless photons*, quanta of the electromagnetic field.) Today, the list (see Table 1) has grown to some thirty-odd particles (including their antiparticles). Faced with such a plethora of

TABLE 1
Elementary particles†

Class	Name	Particle	Anti-particle	Spin parity	Strangeness	~ Rest mass (Mev)	~ Mean life (sec)	Decay modes (ratio %)
Photon	Photon	γ		1	0	0	Stable	
Leptons	Neutrino	ν_e	$\bar{\nu}_e$	$\frac{1}{2}$		$>$0·00025	,,	
		ν_μ	$\bar{\nu}_\mu$	$\frac{1}{2}$		$>$2·5	,,	
	Electron	e^-	e^+	$\frac{1}{2}$		0·510976	,,	
	Muon	μ^-	μ^+	$\frac{1}{2}$		105·655	$2{\cdot}212 \times 10^{-6}$	$e^-\bar{\nu}_e\nu_\mu$ (100)
	Pion	π^+	π^-	0^-	0	139·58	$2{\cdot}547 \times 10^{-8}$	$\mu^+\nu_\mu$ (100)
		π^0		0^-	0	134·97	$1{\cdot}05 \times 10^{-16}$	2γ (98·8), γe^+e^- (1·2)
	Kaon	K^+	K^-	0^-	1	493·98	$1{\cdot}227 \times 10^{-8}$	$\mu^+\nu_\mu$ (64·2)
								$\pi^+\pi^0$ (18·6)
								$\mu^+\pi^0\nu_\mu$ (4·8)
								$e^+\pi^0\nu_e$ (5·0)
								$\pi^+\pi^+\pi^-$ (5·7)
								$\pi^+\pi^0\pi^0$ (1·7)
		K^0	$\bar{K}^0$	0^-	1	497·9	$K_1^0(0{\cdot}90 \times 10^{-10})$	$\pi^+\pi^-$ (69·4)
								$\pi^0\pi^0$ (30·6)
							$K_2^0\ (6{\cdot}3 \times 10^{-8})$	$\pi^+\pi^-\pi^0$ (8·7)
								$3\pi^0$ (38)
								$\pi^\pm e^\mp\nu_e$ (28·3)
								$\pi^\pm\mu^\mp\nu_\mu$ (25·0)
Baryons	Nucleon	p	$\bar{p}$	$\frac{1}{2}^+$	0	938·213	Stable	
		n	$\bar{n}$	$\frac{1}{2}^+$	0	939·507	$1{\cdot}013 \times 10^3$	$pe^-\bar{\nu}_e$ (100)
	Λ Hyperon	Λ^0	$\bar{\Lambda}^0$	$\frac{1}{2}^+$	-1	1115·38	$2{\cdot}57 \times 10^{-10}$	$p\pi^-$ (66), $n\pi^0$ (34)
	Σ ,,	Σ^+	$\bar{\Sigma}^+$	$\frac{1}{2}^+$	-1	1189·40	$0{\cdot}78 \times 10^{-10}$	$p\pi^0$ (51), $n\pi^+$ (49)
		Σ^0	$\bar{\Sigma}^0$	$\frac{1}{2}^+$	-1	1191·5	$10^{-22} < \tau < 10^{-11}$	$\Lambda\gamma$ (100)
		Σ^-	$\bar{\Sigma}^-$	$\frac{1}{2}^+$	-1	1195·96	$1{\cdot}59 \times 10^{-10}$	$n\pi^-$ (100)
	Ξ ,,	Ξ^0	$\bar{\Xi}^0$	?	-2	1314·3	$3{\cdot}06 \times 10^{-10}$	$\Lambda\pi^0$ (100)
		Ξ^-	$\bar{\Xi}^-$	$\frac{1}{2}$	-2	1320·8	$1{\cdot}74 \times 10^{-10}$	$\Lambda\pi^-$ (100)
	Ω ,,	Ω^-	$\bar{\Omega}^-$	$\frac{3}{2}^+$	-3	1675	$0{\cdot}7 \times 10^{-10}$	$\Xi\pi$(?), ΛK(?)

† The antiparticles are supposed to have the same spin, mass and mean life as the particles; strangeness and parity of the antibaryons are of opposite sign to that of the baryons.

"elementary" particles the first reaction of the physicist is to classify them according to some meaningful scheme.

The simplest scheme is to classify the particles according to their masses. Thus, one can group the particles into four classes: *photon*, *leptons*, *mesons*, *baryons*, the last three denoting light, medium and heavy masses respectively. The massless photon is a class by itself, and the neutrinos, although having no mass, are grouped with leptons for reasons to be clarified later. It was soon clear, however, that the particles in each of the classes shared certain properties in common which are characteristic of the class to which they belong.

We can thus also classify the particles according to the types of interactions responsible for the various processes in which the particles participate. Leaving aside the gravitational interaction, whose strength seems to be so far negligible compared to that of the other types, there appears to be roughly three main types of interactions in elementary particle physics. These are the *strong*, *electromagnetic* and the *weak* interactions.

Thus the photon, being the carrier of electromagnetic interaction, interacts electromagnetically with all charged particles and the strength of the interaction is characterized by the fine structure constant $e^2/(\hbar c) = 1/137$. It does not have a weak or strong interaction with any of the other particles. All members of the lepton class interact weakly with each other and with the members of the meson and baryon classes. The relative strength of weak interaction is of the order of 10^{-13}. The members of the meson and baryon classes in general interact strongly with each other and with the members of the lepton class. The relative strength of strong interactions is of the order of unity. And, of course, all charged particles interact electromagnetically with each other.

A study of the processes involving the three types of interactions shows that they are governed by certain conservation laws. For example *all* processes are found to obey laws of conservation of *energy*, *momentum* and *angular momentum* (spin) and *charge number* (that is, the total charge). The conservation laws, in turn, imply certain selection rules; for example, conservation of momentum and energy forbids the decay of massless photon into neutrinos and vice versa. Conservation of charge explains why the electron cannot decay into neutrinos and photons. To ensure the stability of the proton one needs a further conservation law. One, therefore, assigns a so-called *baryon number* $B = +1$ to each member of the baryon class (for antibaryons $B = -1$) and $B = 0$ to the rest of the particles: photon, leptons and mesons. Conservation of baryon number then guarantees the stability of proton. For processes involving leptons one needs an analogous quantum number, the *lepton number* $L = +1$ for each lepton (for antileptons $L = -1$) and $L = 0$ for the rest of the particles: photon, mesons and baryons. The baryon and lepton numbers are also conserved in all three interaction processes.

There exist multiplets of particles having approximately the same mass but different charges and therefore different electromagnetic interactions. One can differentiate each charge state within every such multiplet by assigning to each particle a new quantum number I_3, called the *isospin*. It turns out that the strong interactions that operate between the mesons and baryons are independent of the charge states of the multiplets. The *isospin is, however, conserved only in strong interactions*. It is thus an example of an *approximate* conservation law in contrast to the *absolute* conservation laws of energy-momentum, spin, charge, baryon and lepton numbers and so forth, which are obeyed by all three interactions.

Another example of approximate conservation laws is the conservation of *strangeness quantum number S*, introduced to explain the paradox of strong production and weak decays of K mesons and hyperons. The strangeness number is conserved in strong and electromagnetic interactions but is violated in weak interactions. The electric charge, isospin, baryon and strangeness number are related to each other by the so-called Gell-Mann-Nishijima equation

$$Q = I_3 + \frac{S+B}{2} \tag{260}$$

Finally, we shall have occasion to refer to conservation laws of *parity* and *particle conjugation parity numbers* both of which are violated in weak interactions but are more or less conserved in strong and electromagnetic interactions. It appears that stronger the interaction the more laws of conservation are obeyed by the same.

At present, a theoretical (*albeit semi-quantitative*) description of the spectrum of particles and their properties is provided by the quantum theory of fields, in which one considers the quantized proper field associated with each particle. In Part 2B we considered the electron-photon system from this point of view. The standard formalism of quantum field theory (although highly successful in electrodynamics), when applied to other particle systems have not been so fruitful from the quantitative viewpoint as far as *dynamics* is concerned. However, one fundamental aspect of the theory, namely, *symmetry* has provided us a somewhat deeper understanding of the various conservation laws governing the interactions among the fundamental particles.

We have already seen (in Part 2B and elsewhere) that conservation laws are *implied* by invariance principles. For example, invariance of the action integral under co-ordinate transformations in general relativity and unified field theories leads to conservation identities. Invariance under linear translation and rotation implies conservation of energy-momentum and angular momentum in classical Lorentz invariant field theories. In classical field theory the connection between conservation laws and invariance under continuous groups of transformations *in general* was recognized first by E. Nöther.†

For quantized systems the connection is provided by the following basic theorem.‡

A physical system is said to possess a symmetry or to be invariant under a symmetry operation if there exists a correspondence $|\phi\rangle \to |\hat{\phi}\rangle$ between the ray vectors $|\phi\rangle$ characterizing the original physically realizable states of the system and $|\hat{\phi}\rangle$, the symmetry transformed states, such that all transition probabilities are preserved, that is,

$$|\langle\hat{\phi}|\hat{\psi}\rangle|^2 = |\langle\phi|\psi\rangle|^2 \tag{261}$$

According to the theorem the transformation $|\phi \to |\hat{\phi}\rangle$ characterizing a symmetry operation is "essentially" unique, and this transformation is either unitary or anti-unitary, that is,

$$|\hat{\phi} = S\,|\phi\rangle \tag{262}$$

where S is a unitary or anti-unitary§ operator. The transformation (262) induces a similarity transformation on the observables

$$\alpha \to \hat{\alpha} = S\alpha S^{-1} \tag{263}$$

so that all expectation values are preserved. The symmetry transformation must also preserve the expectation values at *all* times, so that the dynamics of the system remain unaffected. Now, the dynamics of a quantized field system, for example, is governed by the total Lagrangian $\mathscr{L}$. The symmetry transformation (262) would leave the dynamics of the system invariant if

$$\hat{\mathscr{L}} = S\mathscr{L}S^{-1} = \mathscr{L} \quad \text{or} \quad [\mathscr{L}, S] = 0 \tag{264}$$

This usually implies that S also commutes with the total Hamiltonian H of the system

$$[H, S] = 0 \tag{265}$$

In other words S is a constant of motion. If, furthermore, S is a hermetian operator, S is also an observable and thus we obtain a *conserved observable* quantity.

Apart from suggesting "explanations" for the conservation laws, the

† Nöther, E. (1918) *Nachr. Ges. Wiss. Göttingen.*

‡ Wigner, E. P. (1959) "Group Theory and its Applications to the Quantum Mechanics of Atomic Spectra", Academic Press, New York.

§ S is called anti-unitary if (i) S is antilinear, that is, $S(\alpha\,|f\rangle + \beta\,|g\rangle) = \bar{\alpha}S\,|f\rangle + \bar{\beta}S\,|g\rangle$ and (ii) $S^* = S^{-1}$.

notion of symmetry enables us to classify the various particles and predict their qualitative behaviour, even though we are unable to work out their dynamics. The symmetries themselves can be classified into two fundamental groups:

(i) Symmetries of the physical space-time.
(ii) Symmetries of the so-called internal "space".

The two groups of symmetries are independent of each other. Symmetries of the physical space-time consist of a *continuous* group, the *connected inhomogeneous* Lorentz group and the *discrete* space-time transformations. As parts of the connected inhomogeneous Lorentz group the *spatial* rotations enable us to classify all particles into particles of different spins and the space-time translations into particles of different energy-momentum values. The discrete space-time transformations are space reflections leading to the concept of intrinsic parity, and time reflections.

Symmetries of internal "space" include charge conjugation transformation, electromagnetic gauge transformation which leads to the conservation of charge and various other gauge transformations of the first kind leading to the concept of baryon and lepton numbers, isospin and strangeness, and so forth.

The *homogeneous* Lorentz group L is the group of linear homogeneous transformations (μ, $\mu' = 0, 1, 2, 3$)

$$x^{\mu} \to x^{\mu'} = L^{\mu'}_{\mu} x^{\mu} \tag{266}$$

of the space-time co-ordinates† ($x^0 = t, x^1, x^2, x^3$) such that

$$L^{\mu'}_{\nu} g_{\mu'\lambda'} L^{\lambda'}_{\alpha} = g_{\nu\alpha} \tag{267}$$

$$g_{\mu'\lambda'} \equiv g_{\mu\lambda}; \quad -g_{00} = g_{11} = g_{22} = g_{33} = -1, \quad g_{\mu\nu} = 0 \quad \text{for } \mu \neq \nu$$

so that, we have also $\det |L^{\mu'}_{\mu}| = \pm 1$. Moreover, the transformed co-ordinates must be real. Therefore $L^0_0 \geqslant +1$ or $\leqslant -1$. L contains the following four distinct pieces:

$L^{\uparrow}_{+}$: $\det L = +1$, $\quad L^{0'}_{0} \geqslant +1$ containing the Identity

$L^{\uparrow}_{-}$: $\det L = -1$, $\quad L^{0'}_{0} \geqslant +1$ containing P (Parity)

$L^{\downarrow}_{+}$: $\det L = +1$, $\quad L^{0'}_{0} \leqslant -1$ containing T (Time inversion)

$L^{\downarrow}_{-}$: $\det L = -1$, $\quad L^{0'}_{0} \leqslant -1$ containing PT (Space-time inversion)

† Here, and in the following $c = \hbar = 1$.

where

$$
\begin{aligned}
P : x^0 &\to x^{0'} = x^0 \\
x^k &\to x^{k'} = -x^k \\
T : x^0 &\to x^{0'} = -x^0 \\
x^k &\to x^{k'} = x^k
\end{aligned}
$$

Out of these pieces one can form the various *sub-groups* of L

$L_+^\uparrow$, the *restricted proper* Lorentz group

$L^\uparrow = L_+^\uparrow \cup L_-^\uparrow$, the *orthochronous* Lorentz group

$L_+ = L_+^\uparrow \cup L_+^\downarrow$, the *proper* Lorentz group

$L_0 = L_+^\uparrow \cup L_-^\downarrow$, the *orthochorous* Lorentz group

$L_+^\uparrow$ is *connected* in the sense that all its elements can be built up by continuous transformations from the Identity; it is also an invariant subgroup of L.

By adding space-time translations to L we obtain the *inhomogeneous* Lorentz group (also called the *Poincaré group*)

$$\mathscr{L} : x \to x' = Lx + a \tag{268}$$

It contains, as subgroups, L and the *translation group*

$$A : x \to x' = x + a \tag{269}$$

which is an invariant (abelian) subgroup of $\mathscr{L}$, that is, $\mathscr{L}/A \simeq L$. $\mathscr{L}$ also has four distinct pieces $\mathscr{L}_+^\uparrow$, $\mathscr{L}_-^\uparrow$, $\mathscr{L}_+^\downarrow$ and $\mathscr{L}_-^\downarrow$.

It is the *connected inhomogeneous Lorentz group* $\mathscr{L}_+^\uparrow$ which is of interest here. According to (262) the correspondence between state vectors relative to two sets of observers which are related by an element of $\mathscr{L}_+^\uparrow$ is given by

$$|\hat{\phi}\rangle = S(\mathscr{L}_+^\uparrow)\,|\phi\rangle \tag{270}$$

where $S(\mathscr{L}_+^\uparrow)$ forms a unitary (or anti-unitary) representation of $\mathscr{L}_+^\uparrow$ on the Hilbert space of state vectors. The connectivity of $\mathscr{L}_+^\uparrow$ implies that $S(\mathscr{L}_+^\uparrow)$ is a set of unitary operators rather than anti-unitary. One is thus led to the problem of finding unitary representations of $\mathscr{L}_+^\uparrow$. Once $S(\mathscr{L}_+^\uparrow)$ is known one has obtained effectively also the Schrödinger wave equations of the system, because the unitary operator $U(t, t_0)$ in Part 1, equation (144) is simply $S(a \in \mathscr{L}_+^\uparrow)$ where a corresponds to a time translation: $t_0 \to t + t_0$. The classification of all unitary representations of $\mathscr{L}_+^\uparrow$ yields, therefore, also all possible relativistic wave equations. It is thus natural, at this stage, to interpret an *elementary particle system as the manifold of state vectors corresponding to an irreducible representation of* $\mathscr{L}_+^\uparrow$.

$\mathscr{L}_+^\uparrow$ has the following Hermetian infinitesimal generators: (i) p_μ, corresponding to translations along the x^μ co-ordinate, (ii) $M_{\mu\lambda} = -M_{\lambda\mu}$ corresponding to rotations in x^μ-x^λ plane. The commutation relations satisfied by the generators are

$$[M_{\mu\lambda}, p_\nu] = i(g_{\lambda\nu}p_\mu - g_{\mu\nu}p_\lambda) \tag{271}$$

$$[p_\mu, p_\nu] = 0 \tag{272}$$

$$[M_{\mu\lambda}, M_{\alpha\beta}] = -i(g_{\mu\alpha}M_{\lambda\beta} - g_{\lambda\alpha}M_{\mu\beta} + g_{\mu\beta}M_{\alpha\lambda} - g_{\lambda\beta}M_{\alpha\mu}) \tag{273}$$

p_μ and $M_{\mu\lambda}$ generate the Lie algebra associated with $\mathscr{L}_+^\uparrow$. Let

$$\omega_\alpha = \tfrac{1}{2}\varepsilon_{\alpha\beta\gamma\delta}M^{\beta\gamma}p^\delta \tag{274}$$

so that

$$[\omega_\alpha, p_\beta] = 0; \qquad [\omega_\alpha, \omega_\beta] = i\varepsilon_{\alpha\beta\gamma\delta}\omega^\gamma\omega^\delta; \qquad [M_{\mu\nu}, \omega_\alpha] = i(g_{\nu\alpha}\omega_\mu - g_{\mu\alpha}\omega_\nu) \tag{275}$$

Then

$$P = p_\mu p^\mu \tag{276}$$

and

$$W = -\omega_\mu\omega^\mu \tag{277}$$

commute with all the infinitesimal generators and are, therefore, multiples of identity for every irreducible representation of $\mathscr{L}_+^\uparrow$. Their eigenvalues can now be used to characterize an irreducible representation.†

The generators p_μ corresponding to the abelian subgroup A, after exponentiation, provide the unitary operator corresponding to translation by a_μ:

$$S(a) = \exp(-ia_\mu p^\mu) \tag{278}$$

p_μ thus corresponds to the energy-momentum operators of a particle in the Hilbert space of representation. For real particles we need to consider only the cases $p^2 \equiv p_\mu p^\mu = m^2$, a positive constant and $p^2 = 0$. In the former case there are two irreducible representations for each eigenvalue of P and W, one for each value of $p_0/|p_0|$. The eigenvectors can thus be specified as $|\mathbf{p}', \omega'\rangle$. In the rest frame

$$\mathbf{p}' = 0, \qquad p'_0 = m, \qquad W = m^2\mathbf{S}^2$$

where

$$\mathbf{S} \equiv (M_{23}, M_{31}, M_{12})$$

with

$$[S_k, S_l] = i\varepsilon_{klm}S_m \tag{279}$$

† Wigner, E. P. and Bargman, V. (1948) *Proc. natn. Acad. Sci. U.S.A.* **34**, 211.

The S_k have the same commutation relations as that of the real three-dimensional rotation group 0_3. The eigenvalues of $\mathbf{S}^2$ are $s(s+1)$ with $s = 0, \frac{1}{2}, 1$, and so forth. Thus the irreducible representations can be labelled by (m, s), where m and s are to be identified with the restmass and spin values of the elementary particles. Particles of zero restmass and discrete spin (for example, neutrino, photon) are limiting cases of the above representations for $m \to 0$.

The *kinematic* characteristics of an elementary particle are therefore associated with the algebraic properties of $\mathscr{L}_+^\uparrow$. The logical course would be therefore to enlarge the space-time group by incorporating somehow also the symmetries of the "internal" space.

The internal symmetry group which leads to the concept of isospin is SU_2, the classical group of all 2×2 (complex) unitary matrices of determinant one, which incidentally is also the simply connected covering group of real, proper three-dimensional rotations. Invariance of interaction under SU_2 leads to "charge independence" (for example, of proton, neutron) of nuclear forces. To incorporate isospin into the space-time group, one can, for example, consider simply the direct product $\mathscr{L}_+^\uparrow \times SU_2$, so that the elementary particle state vectors now become elements of $\mathscr{H}(\mathscr{L}_+^\uparrow) \otimes V(SU_2)$, where $\mathscr{H}(\mathscr{L}_+^\uparrow)$ is the Hilbert space representation (irreducible) of $\mathscr{L}_+^\uparrow$, and $V(SU_2)$ is a vector space representation (irreducible) of SU_2.

Gell-Mann and Ne'eman† have shown that the appropriate internal symmetry group which incorporates both isospin and hypercharge‡ is SU_3 and they succeeded in associating eight baryons ($p, n, \Xi^0, \Xi^-, \Sigma^+, \Sigma^0, \Sigma^-, \Lambda^0$) and eight mesons ($K^+, K^0, \bar{K}^0, \bar{K}^-, \Pi^+, \Pi^0, \Pi^-, \eta$) with the eight dimensional representation of SU_3.

It would therefore be desirable to combine not SU_2 but SU_3 with $\mathscr{L}_+^\uparrow$. This has proven to be not an easy problem. Sakita, Gursey and Radicati§ succeeded in combining only the *static* spin part of $\mathscr{L}_+^\uparrow$ with SU_3. For a fixed momentum the spin of a particle is associated with irreducible representations of SU_2. The group which combines the internal symmetry group SU_3 and the static spin part of the space-time group $\mathscr{L}_+^\uparrow$, that is, SU_2, is found to be SU_6 which contains $SU_3 \times SU_2$ as a subgroup. SU_6 mixes the ordinary spin and SU_3 "spin" co-ordinates in such a way that particles with different spin as well as different isospin and strangeness may be in a same super-multiplet.

Attempts to make SU_6 relativistic (non-trivially) have, however, led to an impasse. A fundamental theorem due to McGlinn|| shows that any fusion of

† Gell-Mann, M. and Ne'eman, Y. (1964) "The Eightfold Way", Benjamin, New York.

‡ Hypercharge $Y = S+B$ = Strangeness + Baryon number.

§ Sakita, B. *Phys. Rev.* **136** B, 1756; Gursey, F. and Radicati, L. A. (1964) *Phys. Rev. Lett.* **13**, 173.

|| McGlinn, W. D. (1964) *Phys. Rev. Lett.* **12**, 467.

the internal and space-time symmetry groups under "reasonable" hypotheses is necessarily a trivial one. Suppose that the relevant internal symmetry is I (for example, SU_3). If we take for the complete symmetry group simply the direct product $\mathscr{L}_+^\uparrow \times I$ it would follow that particles belonging to the same irreducible representation of I would all have the same mass because I_k, the generators of I would commute with p_μ, the translation operators. The problem is therefore to find a Lie Group G which has as generators those of $\mathscr{L}_+^\uparrow$ and I, but for which $[I_k, p_\mu] \neq 0$, so that, particles belonging to the same representation of I may have different masses. On the other hand, for obvious physical reasons, it is also necessary to assume that the quantum numbers associated with I be invariant under homogeneous Lorentz transformations, that is, I_k commute with the generators of the *homogeneous* Lorentz transformations. McGlinn shows that this automatically implies that $[I_k, p_\mu] = 0$ and consequently $G = \mathscr{L}_+^\uparrow \times I$.

The task of unifying the "internal" symmetry group with the "external" space-time group remains one of the central problems in elementary particle physics. Symmetry analysis, however, can clarify only one albeit important aspect of what must be the future comprehensive theory of "all" matter.

Heisenberg's unified field theory. A concerted attempt to unify all elementary particles in a single formalism based more or less on the existing ideas of quantum field theory has been made by Heisenberg *et al.*† We have already mentioned that in conventional field theory one starts by assigning a field operator to each elementary particle. This procedure implicitly assumes the existence of a criterion by means of which we can distinguish an elementary particle from a compound system. Heisenberg believes that it is essential to recognize that no such criterion exists. However, this need not prevent us from using the term "elementary particle" whenever we want to disregard internal structure of a subsystem.

We have also seen that conventional prescription of quantization when applied to Lorentz-invariant interacting *local* field (that is, differential) equations leads to divergent quantities. In order to obtain convergent results one has therefore been forced to introduce the interaction essentially as a *non-local* one, for example, by means of a so-called cut-off procedure. This, however, implies deviation from the kind of causality that follows from the space-time structure of special relativity. It would seem that complete local causality is incompatible with quantization. Therefore any field theory of elementary particles must try to resolve the central mathematical problem of combining quantization with a certain greater or lesser degree of relativistic causality.

In order to avoid these difficulties many physicists have abandoned the concept of fields altogether and have concentrated on a theory solely of

† See Reference in Introduction.

S-matrices, which are immediately related to observable parameters in any collision process. It is possible to construct S-matrices compatible with the requirements of unitarity and special relativity without encountering divergent terms. Together with postulates of analyticity† and crossing symmetry the S-matrix formalism has been very useful in analyzing cross-sections of collision processes. It cannot be considered, however, a complete theory of matter; it bypasses some of the fundamental problems mentioned earlier. What is needed of course is a comprehensive theoretical formalism that allows one to calculate the masses of particles and at the same time the S-matrix from the fundamental equations of the theory. According to Heisenberg‡ the basic requirements for such a theory should be as follows.

1. *The field operators necessary for formulating the fundamental equations should not refer to any specific particle like proton or meson and so forth, but should refer to matter in general.*
2. *The particles (elementary or compound) should be derived as eigen solutions of the fundamental field-equation.*
3. *The fundamental field-equation must be non-linear in order to represent interaction. The masses of the particles should be a consequence of this interaction. The concept of a "bare particle" therefore has no meaning.*
4. *Selection rules for creation and decay of particles should follow from symmetry properties of the fundamental equation. Therefore, the empirical selection rules should provide the most detailed information on the structure of the equation.*

After several heuristic attempts Heisenberg was led finally to the non-linear spinor equation§

$$\gamma_\mu \frac{\partial \psi}{\partial x_\mu} \pm l^2 \gamma_\mu \gamma_5 \psi (\bar{\psi} \gamma_\mu \gamma_5 \psi) = 0 \tag{280}$$

for the elementary matter field ψ. Here l is a fundamental physical constant; no other physical constants (besides c and $\hbar$) are to appear in the theory.

The equation (280) is invariant under $\mathscr{L}_+^\uparrow$. It is therefore possible to derive conservation laws for energy, momentum and angular momentum in both the classical and quantum formalism.

The equation is also invariant under the Pauli-Gürsey transformations‖:

$$\psi \to a\psi + \gamma_5 C^{-1} \bar{\psi}^T; \qquad \bar{\psi} \to a^* \bar{\psi} + {}^*\psi^T C \gamma_5 \tag{281}$$

† Chew, G. F. (1962) "S-matrix Theory of Strong Interactions", Benjamin, New York.
‡ Heisenberg, W. (1957) *Rev. Mod. Phys.* **29**, 269.
§ ψ, $\bar{\psi}$, γ_μ are as in Dirac's equation and $\gamma_5 = \gamma_1\gamma_2\gamma_3\gamma_4$.
‖ Pauli, W. (1957) *Nuovo Cim.* **6**, 204; Gürsey, F. (1958) *Nuovo Cim.* **7**, 411.

with
$$|a^2|+|b^2| = 1; \quad C\gamma_\mu C^{-1} = -\gamma_\mu^T; \quad C\gamma_5 C^{-1} = \gamma_5^T; \quad C^T = -C$$

and the Touschek transformation†:
$$\psi \to e^{i\alpha\gamma_5}\psi; \quad \bar\psi \to \bar\psi e^{i\alpha\gamma_5} \tag{282}$$

The transformations (281) constitute a group which is isomorphic to the 3-dimensional rotation group. The corresponding generators I_1, I_2, I_3 are to be identified with the three components of isospin.

The transformations (282) also form a group which commutes with (281). The corresponding generator I_N is related to the Baryon number.

Equations (281) and (282) combined is isomorphic to 2×2 (complex) unitary matrices U_2. The sub-group (281) then corresponds to SU_2.

In the quantum theory one therefore gets three independent quantum numbers I, I_3, I_N which can take half-integral or integral values. These quantum numbers are, however, not enough for an empirical classification of the elementary particle system. One introduces the charge Q and the Baryon number B by

$$Q = I_3 + l_Q/2 \tag{283}$$
$$B = I_N + \frac{l_N}{2} \tag{284}$$

where l_Q and l_N are two more quantum numbers which are related to the strangeness S by

$$S = l_Q - l_N \tag{285}$$

l_Q and l_N then must be able to take positive or negative values whereas S need not take all positive and negative integral values but simply $0, \pm 1, 2$.

TABLE 2

	p	n	$\bar n$	$\bar p$	e^+	$\bar\nu$	ν	e^-	π^+	π^0	π^-	Λ^0	Σ^+	Σ^0	Σ^-	Ξ^-	Ξ^0	K^+	K_0	$\bar K_0$	K^-	μ^+	μ^-	γ
I_3	$\frac12$	$-\frac12$	$\frac12$	$-\frac12$	$\frac12$	$-\frac12$	$\frac12$	$-\frac12$	1	0	-1	0	1	0	-1	$-\frac12$	$\frac12$	$\frac12$	$-\frac12$	$\frac12$	$-\frac12$	0	0	0
$l_Q/2$	$\frac12$	$\frac12$	$-\frac12$	$-\frac12$	$\frac12$	$\frac12$	$-\frac12$	$-\frac12$	0	0	0	0	0	0	0	$-\frac12$	$-\frac12$	$\frac12$	$\frac12$	$-\frac12$	$-\frac12$	1	-1	0
Q	1	0	0	-1	1	0	0	-1	1	0	-1	0	1	0	-1	-1	0	1	0	0	-1	1	-1	0
I_N	$\frac12$	$\frac12$	$-\frac12$	$-\frac12$	$-\frac12$	$-\frac12$	$\frac12$	$\frac12$	0	0	0	$\frac12$	$\frac12$	$\frac12$	$\frac12$	$\frac12$	$\frac12$	0	0	0	0	$-\frac12$	$\frac12$	0
$l_N/2$	$\frac12$	$\frac12$	$-\frac12$	$-\frac12$	$\frac12$	$\frac12$	$-\frac12$	$-\frac12$	0	0	0	$\frac12$	$\frac12$	$\frac12$	$\frac12$	$\frac12$	$\frac12$	0	0	0	0	$\frac12$	$-\frac12$	0
B	1	1	-1	-1	0	0	0	0	0	0	0	1	1	1	1	1	1	0	0	0	0	0	0	0
$l_Q - l_N$	0	0	0	0	0	0	0	0	0	0	0	-1	-1	-1	-1	-2	-2	1	1	-1	-1	1	-1	0

Reproduced from *Zeitschrift für Naturforschung* with permission.

It is therefore necessary to have at least one more continuous one-parameter group to account for l_Q or l_N, whereas S could be accounted for by a discrete group. In fact (280) is also invariant under the following transformation:

$$\psi \to \eta^{3/2}\psi(x\eta, l\eta); \quad \eta \text{ real} \tag{286}$$

† Touschek, B. (1957) *Nuovo Cim.* **5**, 1281.

Here ψ is to be regarded as a function of x and l. The transformation (286) commutes with all elements of the homogeneous Lorentz group, of (281) and (282) but not with space-time translations. This means that in the quantum theory it is not possible to characterize a particle-state simultaneously by the mass eigenvalue and the quantum number associated with (286). This turns out to be irrelevant because the mass eigenvalues behave like Z/l, where the number Z is a characteristic of the state. Since it is only the ratio of mass eigenvalues which is relevant, the important quantity is Z (which is determined by the eigen solutions of (280)) and not Z/l. In other words, if, instead of the usual way, we define space-time translations as follows

$$\psi(x_\mu, l) \rightarrow \psi(x_\mu + a_\mu l, l) \tag{287}$$

then (286) commutes with (287). In quantum theory there exists then an operator 0_η with the property

$$0_\eta \psi(x, l) 0_\eta^{-1} = \eta^{3/2} \psi(x\eta, l\eta) \tag{288}$$

The transformation (288) is not unitary. It can be, however, made pseudo-unitary in a Hilbert space with indefinite metric. Under appropriate assumptions 0_η can lead to the quantum number $\frac{l_N}{2}$ (or $\frac{l_Q}{2}$ for particles with $S = 0$).

So far we have considered only continuous groups. Equation (280) is invariant under the usual discrete operations: Parity P, charge conjugation C and time-reversal T. The operations P, C and T, however, commute only in parts with the continuous groups. This difficulty can be removed by defining modified P and C operations which are closely related to the usual ones.

Equation (280) is also invariant under the discrete transformation

$$\psi(x, l) \rightarrow \psi(x, -l) \tag{289}$$

which commutes with all the other transformations and serves to define a further quantum number. It is not yet clear whether the strangeness number S should be identified with it; the above list of discrete symmetrics of (280) is not exhaustive.

As in the linear theory, the principle of microcausality is assumed to be valid also here. That is, commutator (or anticommutator) of field operators between points separated by space-like distances vanish. In the linear case this, together with the differential equations for the field operators are sufficient to determine completely the quantization process; the commutators are at the same time solutions of the field equations and one therefore obtains δ (or δ')-type singularities on the light cone. It is to be expected that the principle of microcausality would also (together with the field equations) determine completely the quantization in the non-linear case except that, because of the non-linear differential equation, the singularities on the light cone would be markedly different from the δ-type. The principle of micro-causality is compatible with the symmetry properties of (280) and therefore the quantization process does not destroy any symmetry. It is by no means certain, however, that there exists a vacuum state possessing the complete symmetry of the field-equations. If it is impossible to construct such a vacuum state it would mean that one would have instead a highly degenerate ground-state from which all elementary particles could be realized. This groundstate may have, for example, infinite isospin; the strange particles could then be interpreted as states which "borrow" an isospin $\frac{1}{2}$ or 1 from the groundstate.

The groundstate expectation values of simple products of only two field operators may be expected to have the complete symmetry and thus the form

$$\langle 0'|\psi_\alpha(x)\bar{\psi}_\beta(x')|0'\rangle$$
$$= l^{-3}\, 1/(2\pi)^4 \int_0^\infty \rho(\zeta)\, d\zeta \int dq \exp\{iq(x-x')/l\} \frac{q_\nu(\gamma_\nu)_{\alpha\beta}}{q^2+\zeta} \tag{290}$$

where the mass-spectrum function $\rho(\zeta)$ is to be determined by (208) whose non-linearity precludes any δ or δ'-functions on the light cone. Therefore

$$\int \rho(\zeta)\, d\zeta = 0; \qquad \int \zeta\rho(\zeta)\, d\zeta = 0 \tag{291}$$

so that $\rho(\zeta)$ must take positive and negative values which is possible only in the framework of a Hilbert space with an indefinite metric.

It is well known that four complex field operators satisfying a linear first-order differential equation can describe only two spinor particles. For example, the four component Dirac's spinor equation can describe, say, free electrons and positrons. Equation (280), is, however, non-linear, albeit of first order, and therefore the question of how many different kinds of particles can be described by it cannot be answered by a rule of thumb. What is important is the group (that is, symmetry) property of the field operators and their irreducible components. The totality of matrix elements of $\psi(x)$ would contain a series of irreducible components which transform differently under $\mathscr{L}_+^\uparrow$, (281), (282) and (286), (289) and so forth. In principle, therefore, one can obtain many types of stable particle states as well as unstable ones.

The matrix elements $\varphi(x) \equiv \langle 0'|\psi(x)|p\rangle$ *of* $\psi(x)$ *between the groundstate* $|0'\rangle$ *and one-particle states* $|p\rangle$ *can be interpreted in a certain sense as the wave functions of the corresponding elementary particles, which then satisfy free-field linear wave equations.* The Tamm-Dancoff method together with an approximation procedure is used to derive these linear one-particle wave equations from the original non-linear equation (280). The *one-particle wave equations then furnish estimates of particles masses.* Although the masses and symmetry properties of nucleons and mesons seem to agree qualitatively with experimental values it is difficult to judge the use and validity of the approximation procedure employed as it has been possible to work out only the lowest order of approximation (higher orders of approximation being exceedingly complicated).

It turns out that fermions of non-zero mass are described by wave functions satisfying *second order Klein-Gordon spinor* equations instead of the usual Dirac equation.

The problem of scattering of, for example, π-mesons by nucleons can be treated by a method related more closely to the Bethe-Salpeter formalism than to the Tamm-Dancoff method. The theory leads to a relativistic pseudovector coupling as the main interaction term and to a value of the coupling constant of the right order of magnitude.

The analysis of β-decay shows that $c_S = c_T = c_P = 0$, and, in the lowest approximation $c_A = -c_V$; in the higher approximation the ratio c_A/c_V is somewhat altered.

In principle, therefore, the masses of the elementary particles and the coupling constants of various interaction processes, as well as scattering amplitudes could be calculated from the theory.

Conclusion

The concept of fields as primary objects of physical reality would therefore seem to have served a very fruitful purpose in the development of the fundamental theories of physics. The notion of a field is obviously intimately connected with our concept of space-time as a basically simple geometric manifold. Should it be necessary to modify the existing concept of space-time the field concept would also have to undergo a corresponding change.

We shall simply conclude by pointing out a fundamental difficulty associated with the field concept.

In quantum electrodynamics one makes the tacit assumption that the electromagnetic field operators $f_{\mu\lambda}(x)$ attached to a space-time point x are *observable* and therefore, in principle, precisely *measurable* at x. In quantum theory of a particle one makes a similar assumption that a single observable, for example, the position operator $q(t)$ is, in principle, precisely measurable at any given time t; although in practice one may always deal with a wave packet the theory sets no (non-zero) lower limit to the extension of the wave packet. The question then arises whether there are any limitations, theoretical or otherwise, to the precise measurement of fields.

We note first of all a theoretical dilemma as far as the non-dualistic field theories are concerned, in that, the only way to measure the field strength at a point is to measure the impulse imparted to a test particle; so that it is necessary to have a test particle before a field can be said to exist *operationally*.

The question of measurability of electromagnetic fields has been dealt with by Bohr and Rosenfeld.† They show that as long as we use test bodies of sufficiently large mass, charge and current density spread over a region of space-time and *disregard the limitations due to the atomic structure of such test-bodies*, quantum electrodynamics, as far as the question of measurability is concerned, is a fully consistent, albeit an idealized theory. In other words, fields can be measured over a finite space-time region and the present theory, in principle, sets no lower limit to the size of such regions. However, the concept of localizable fields as an idealization need not be valid in a future comprehensive theory which takes into account of the limitations due to the microscopic structure of test bodies.

† Bohr, N. and Rosenfeld, L. (1933) *Det. Kgl. Danske Videnskab. Selskab. Math.-Fysiske Medd.* **12**, No. 8.

General Reference

Part 1

Barut, A. O. (1964). "Electrodynamics and Classical Theory of Fields and Particles." Macmillan, New York.

Pauli, W. (1958). "Theory of Relativity." Pergamon Press, New York.

Infeld, L. and Plebanski, J. (1960). "Motion and Relativity." Pergamon Press, New York; PWN, Warsaw.

Dirac, P. A. M. (1947). "Principles of Quantum Mechanics." 4th ed. Oxford University Press, London.

Part 2

Rohrlich, F. (1965). "Classical Charged Particles." Addison-Wesley, New York.

Schweber, S. (1961). "Relativistic Quantum Field Theory." Row, Peterson, New York.

Jauch, J. M. and Rohrlich, F. (1955). "Theory of Photons and Electrons." Addison-Wesley, New York.

Part 3

Tonnelat, M. A. (1955). "La Théorie du Champ unifié d'Einstein." Gauthier-Villars, Paris.

Wheeler, J. A. (1962). "Geometrodynamics." Academic Press, New York.

Marshak, R. E. and Sudarshan, E. C. G. (1961). "Introduction to Elementary Particle Physics." Interscience, New York.

Subject Index

G

H

I

K

L

Author Index